Eros and Touch from a Pagan Perspective

Within the past twenty years, contemporary Pagan leaders, progressive Christian and Goddess theologians, advocates for queer and BDSM communities, and therapeutic bodyworkers have all begun to speak forcefully about the sacredness of the body and of touch. Many assert that the erotic is a divinely transformative force, both for personal development and for social change. Although "the erotic" includes sexuality, it is not limited to it; access to connected nonsexual touch is as profound a need as that for sexual freedom and health. In this book, Christine Hoff Kraemer brings together an academic background in religious studies and theology with lived experience as a professional bodyworker and contemporary Pagan practitioner. Arguing that the erotic is a powerful moral force that can ground a system of ethics, Kraemer integrates approaches from queer theology, therapeutic bodywork, and sexual minority advocacy into a contemporary Pagan religious framework. Addressing itself to liberal religious people of many faiths, *Eros and Touch from a Pagan Perspective* approaches the right to pleasure as a social justice issue and proposes a sacramental practice of mindful, consensual touch.

Christine Hoff Kraemer holds a Ph.D. in Religious and Theological Studies from Boston University. She is an instructor at Cherry Hill Seminary and Managing Editor of the Patheos.com Pagan channel. Kraemer's research focuses on contemporary Paganism, popular culture, body theology, and sexuality, and her publications include *Seeking the Mystery: An Introduction to Pagan Theologies* (2012) and *Graven Images: Religion in Comic Books and Graphic Novels* (2010, co-edited with A. David Lewis).

Eros and Touch from a Pagan Perspective

Divided for Love's Sake

CHRISTINE HOFF KRAEMER

Routledge
Taylor & Francis Group
NEW YORK AND LONDON

First edition published 2014
by Routledge
711 Third Avenue, New York, NY 10017

and by Routledge
2 Park Square, Milton Park, Abingdon, Oxon OX14 4RN

Routledge is an imprint of the Taylor & Francis Group, an informa business

Library of Congress Cataloging in Publication Data
Kraemer, Christine Hoff.
Eros and touch from a pagan perspective: divided for love's sake/
Christine Hoff Kraemer.
pages cm
Includes bibliographical references and index.
1. Touch—Religious apects. 2. Senses and sensation—Religious aspects.
3. Human body—Religious aspects. I. Title.
BL53.K69 2013
299'.94—dc23
2013013513

ISBN: 978-0-415-82018-9 (hbk)
ISBN: 978-0-203-63152-2 (ebk)

Typeset in Minion
by Florence Production Ltd, Stoodleigh, Devon, UK

MIX
Paper from responsible sources
FSC www.fsc.org FSC® C013604

Printed and bound in Great Britain by
CPI Group (UK) Ltd, Croydon, CR0 4YY

for my teachers
and those who taught them

Contents

Acknowledgments

The assistance of colleagues, friends, and family has been of enormous benefit in my writing process.

Emily Taylor Merriman was a tireless reader of my early drafts; Kerry Wendt, Sierra Black, and Michael Cohn provided important feedback, and Sarah Twichell and Niki Whiting offered invaluable editing support during the final revision. My appreciation also goes to Ken Irwin and Scott Martin for their expertise as research librarians; Chas Clifton, Nikki Bado, and Wendy Griffin for their encouragement; Steve Wiggins for his facilitation of the publication process; and Matt Major, for the nigh infinite patience it must take to live with an aspiring Pagan theologian.

I have had many valuable teachers in the course of my life, more than I have space to name here. The three threads of religious studies, spiritual practice, and bodywork that came together into this book, however, can be traced most directly to Peter S. Hawkins, T. Thorn Coyle, and David Lauterstein. All are consummate instructors and a source of inspiration to their students, and I have been blessed by their influence on my thinking and my living.

Special thanks go to Alicia Belcher, whose perfectly timed gift inspired me to take a necessary risk. But most importantly of all, my deepest gratitude goes to Nicholas Haasch: without his generous support, this book might never have been written.

Introduction

PAGAN EROTIC THEOLOGY

Up late, reading, writing, a breeze
rifles the pages of my books,
sends my notes flying, my work,
my lifework, onto the floor. This nook,
this cranny where my eyes strain at
lines of cramped type confines my soul
to teetering stacks of dusty
volumes, my cracking skin the toll
I pay for caressing paper
instead of flesh.
This mouth of mine
is less for words than for kissing,
my hands less for scribbling of rhymes
than for finding the body's own
rhythm. But I have grown dry
like the pages of my books. What
is all this struggling for, if I
cannot be juicily human?

"Poem Written in a Carrell," 2005

This book describes a theology of touch, and it comes out of my own experience of healing. It is a different book than I might have written had I come into adulthood in the 1970s rather than in the late 1990s. My mentors in the Pagan community, many of them women of my mother's generation, have spoken to me of being raised with a pervasive body shame that their rituals and books sought to heal. The feminist movement has not eradicated body-hatred from our culture, but it has undoubtedly had its successes. I was one of a generation of young women who grew up reading feminist, body-positive books like *Our Bodies, Ourselves*, and, despite my parents' conservative religious upbringing, they taught me that sexuality was a holy thing and did not condemn me when I made choices different from theirs. Today, when I speak to women younger than myself about classic 1970s and 1980s texts of feminist spirituality, some find it difficult to relate to them: the feelings of shame, powerlessness, and rage expressed in those texts are alien to their experience. Though the problems addressed by second-wave feminism have not disappeared, a shift has occurred. For some young people, learning to recognize the holiness of the body no longer begins from a place of trauma.

Because of the passionate struggle of women and men in the generations before me, I was never taught to be ashamed of my body. Yet a lack of shame is not sufficient for health. The healing I have done, and that I continue to do, was of a more subtle, chronic, and pervasive illness than body-hatred: I did not know how to live in my body, embrace it, feel that I *was* it. I chose an academic career path, and the conditions of constant study worsened the problem. I spent my days reading, writing, and speaking—often *about* embodiment! Yet based on my actions, my value as a person seemed concentrated only in the brain that thought, the mouth that spoke, and the fingers that wrote on chalkboards and typed at keyboards.

Gil Hedley, who, like me, is a bodyworker with a Ph.D. in religion, writes that during his most dedicated periods of academic study, he tended to bodily needs only because they were necessary for him to keep reading.[1] Although I was aware that I was abusing my body during my graduate studies, the constant pressure (both internal and external) to know more, write more, take one more class, or attend just one more conference kept me hunched over computers and books when I could have been exercising, dancing, or lying in the sun. The subject matter of my dissertation only exacerbated the sense of disembodiment. I was studying literature and films by sexual minorities who portrayed their sexuality as sacred, yet the art I analyzed only reminded me of the physical alienation of the life I was leading, a kind of energetic dryness that I tried unsuccessfully to balance with spiritual practices and sexual experimentation. In the end, it wasn't enough. My husband and I brought our marriage to an amicable end; I finished my Ph.D.; and then I packed my belongings, moved across the country, and enrolled in massage school.

Learning to be a massage therapist is not just about studying techniques to release muscle tension. The massage classroom became a safe space for me to experience my body and begin to undo the damage that years of unremitting stress had wrought. I gave and received as many as three massages a week. They were amateur massages from my classmates, but they were more than enough to experience the transformative effects of lessened anxiety. I learned that when I lay my hands on a client, it is not just her body I am contacting, but her entire life history; and conversely, I do not touch her just with my hands, but with my entire history as well. Without being healed myself, I could not help others to heal.

I had been a practitioner of contemporary Paganism for over ten years at that point, and I had many times led rituals to help participants come into the presence of divinity. Giving massage was even more intimate. I learned and saw firsthand how my physical pain, grief, and anxiety affected my ability to give an effective treatment. As a priestess, I had always tried to enter ritual grounded, centered, and calm so that I could create a safe space for others. Massage demanded a whole new level of alignment and self-acceptance. To coax another's body to relax, I had to relax my own; to give another an opportunity to release lingering emotions—as happens not infrequently on the bodywork table—I had to be willing to let my own go. Though my theology had always affirmed the sacredness of my body, more than ever before, being trained as a massage therapist forced me to walk my talk. In the face of the human body's glorious complexity, it taught me humility; in being present for others' healing, it taught me reverence; and in the changes it demanded of me, I learned that to serve others, I first had to care for myself.

Healing is not a process I finished in massage school, nor do I expect it will ever be entirely complete. Yet during that training, I confronted how profoundly my own mental, physical, emotional, and spiritual health impacts those around me, especially those I touch. Pagan priestess Francesca De Grandis writes that "A healthy priest(ess) makes all things sound"[2]—but I was not yet healthy. My life was out of balance, tipped much too far toward intellectual pursuits. Only by stepping away from academia for the better part of a year to concentrate on the care of bodies—others' and my own—did I begin to move toward health. Although while I was in graduate school, this book had already begun to whisper itself to me between the lines of my dissertation, I was not yet the person who could write it.

I have not achieved perfect health, but I have become better able to support others in their own healing. One client, a young Haitian-American woman who had never received massage before, affirmed my work more than she knew when she told me, "That was like—what's the word?—a *ritual*." One might think that, as a massage therapist, I would feel valued only for my hands and arms, much as in academia I felt valued solely for my fingers and brain. But part of the illness that afflicted me in graduate school stemmed from my inability to bring my whole self to my work. Academia may have not-so-subtly encouraged me to leave my body behind, but I was the one who actually distanced myself from my own physicality. The integration and divine connection that I sought in my religious practice could not manifest itself fully until I welcomed my body into my work. Though we stood under fluorescent lights in a sterile medical office, my client called the massage she received "ritual." That day, she affirmed my faith that any touch can be made sacred, and that striving for my own healing also helps others in seeking their own. In learning to bring my whole self—my whole history—to the treatment room, I learned also to bring it to my teaching and writing. My actions, my thoughts, my way of being in the world resonate through all parts of my life, affecting those around me far more deeply than I consciously acknowledge.

Like many contemporary Pagans, my religious beliefs are *panentheistic*—by which I mean, I experience divinity within all things, but also transcending them; divinity is always more than simply the sum of the material world. In a panentheistic model, no person or thing is truly autonomous. All of Being is part of an interdependent and connected system, with each component part affecting the whole. Many readers may have heard of the "Butterfly Effect," based on the principle of sensitive dependence on initial conditions.[3] In this paradigm, the flapping of a butterfly's wings, given a certain set of initial conditions, could trigger a hurricane 2000 miles away: a tiny change in one part of the system can be amplified into massive changes in another. Yet one need not resort to the complexities of chaos science to affirm the concept of interdependence. Ecology, sociology, and other branches of research focus on studying the patterns that arise in ecosystems and in human societies—patterns that, in turn, affect all those individuals within them. My discovery that my health is interdependent with the health of others—particularly when I am acting as a massage therapist or facilitating a ritual, but also in every ordinary moment—is one piece of my lived experience of interdependence.

Although caring for myself also has a positive impact on others, fundamentally I understand care—and more specifically, pleasure—as a holy end in itself. Mutual pleasure, sexual or simply sensual, increases love, respect, and understanding. Skin on skin communicates on a level that words alone cannot reach. For me, and for others in my religious community, interdependence is not merely an intellectual belief, but an ecstatic state of connection to which one can become attuned. All of Being erotically interpenetrates—rain soaking soil, hand clasping hand, breath drawn into lungs, tongue thrust into mouth. To Pagan eyes, the long line of the sky pressed against the earth at the horizon is an intimate sight, one that we reflect when we lie down with our lovers. It is this quality of existence that many Pagans call "Goddess" or "the nature of the gods."

Pagans are not alone in experiencing divinity with the intimacy and sensuality that our culture commonly only associates with sex—although, as I'll argue later, "the erotic" encompasses a great deal more than just sexuality. My thinking about how the erotic and social justice intersect has been deeply affected by queer Christian liberation theologian Marvin M. Ellison. In *Erotic Justice*, Ellison writes:

> At its best Christian moral life is about sharing power. Empowerment means aligning ourselves with that sacred power flowing through breasts, penises, clitorises, naked skin, tongues, fingers, and sweaty, quivering flesh. That power piques people's desire for wholeness, integrity, and at-one-ness with others and the created order. Sexuality and spirituality are unavoidably intertwined as life-giving desire for sensuous connection and radical communion. Sexual passion entices us to become vulnerable and to open ourselves to affirmation and care. Erotic desire energizes and keeps us in touch with that awesome power that moves our hearts and yet also, magnificently, turns the galaxies and rotates the stars above.[4]

In this passage, I hear the echo of my own experience—loving and desirous contact between individuals is the same power that binds us together in group solidarity, and it is also the power that moves the forces of Nature itself. Whether at an individual, social, or cosmic level, the erotic binds us together.

This book offers an erotic theology and ethics not just to Pagans, but to all progressive religious people seeking to embrace their own embodiment. I put it forward not as a prescriptive or creedal system, but rather as one example of a religious approach to eros and touch. Because Pagan rituals and practices often attempt to sacralize the physical and bring participants into close contact, Pagan communities are actively engaged in testing and clarifying new systems of ethics around touch. As a practitioner of a body-centered Paganism and a participant in this process of creating alternative community norms, I feel that contemporary Pagans have much to offer ongoing interfaith conversations about the erotic. In both their triumphs and their failures, Pagans' willingness to question standard touch taboos can teach us a great deal about wider cultural struggles over the body and sexuality.

Defining Terms

This book aims to reach a diverse audience, including those with backgrounds in queer and lesbian, gay, bisexual, transgender (LGBT) studies; Pagan studies; and progressive theology from multiple faiths. For those with knowledge of queer and LGBT studies, I have provided contextualizing information about Paganism as a new religious movement and a skeletal description of theology as a discipline. For those with a background in religious studies in general or Pagan studies in particular, I have drawn some distinctions between religious studies and theology and also offered material related to sexual ethics, consent, and the politics of sexual minority communities. Those familiar with progressive theology are likely to be most familiar with the strategies employed here, but they may still need both the material on the Pagan movement and on queer studies for context. Accordingly, much of this introduction is devoted to the definition of terms. The reader is invited to read this section carefully, skim it, or skip it entirely depending on her needs.

A note on the use of gendered words: When speaking of a person of indeterminate gender, such as "the reader," I've chosen to alternate between "he" and "she," rather than using "he" as if it were gender-neutral or employing the frequently clunky "he or she." For the purposes of this book, I've reserved gender-neutral pronoun neologisms, such as "s/he" and "hir," solely for third-gendered or transgender persons who do not identify as either male or female. Further, to reflect the diversity of Pagan theological attitudes, I have chosen to employ a wide variety of terms for deity rather than choosing just one: some singular, some plural; some singly gendered, others multiply- or indeterminately gendered. I hope this convention helps to convey the dynamic and constantly evolving nature of Pagan perceptions of divinity.

Introducing Contemporary Paganism

This book promises "a Pagan perspective"—and it is indeed just one Pagan perspective, though one that represents significant trends in my religious movement. Due to its diversity and the passionate individualism of many of its practitioners, contemporary Paganism can be a difficult movement to define. Rather than being a single religious tradition, Paganism is instead an umbrella term for a wide variety of loosely associated religious paths. Current estimates put the number of American Pagans at 1.2 million[5], with other significant populations around the world. Of the many Pagan traditions, Wicca is the largest group, and Wicca itself is subdivided into many different groups, paths, and lineages. Other Pagan traditions include

Asatru or Heathenry (Northern European Paganism), Druidry, various kinds of feminist Goddess worship, a variety of non-Wiccan traditions of religious witchcraft, and many reconstructionist groups (who attempt to accurately recreate the practices of ancient religions, such as those of Greece and Egypt). Pagans tend to be innovative and experimental, while also drawing heavily on pre-Christian and indigenous religious traditions for inspiration.

Contemporary Paganism is in the process of formulating its theology, or rather, theologies. The nature of Paganism, however, means that it will probably never have a systematic theology in the way that Christian theologians have striven for. Pagan groups are held together more by shared practice than by shared belief, and Pagans tend to use personal experience as the ultimate basis for their theologies. This results in a certain mutability and context-dependence that is an essential component of the theology. If Pagan theologies maintain their integrity, they will always be in process, constantly reorienting themselves to experiential mysteries that cannot be contained in words. Because life is full of paradox and there is always more data than human beings can effectively evaluate, a certain amount of blurriness when it comes to understanding the divine is part of Paganism's wisdom. I approach writing Pagan theology not as the creation of doctrine, but rather as a dynamic process: I articulate beliefs based on my experience (individual and in community); these beliefs inform my behavior and religious practice; further life experiences alter and refine my beliefs; and so on. Theology itself is a practice, something I *do*, not a mere intellectual exercise.

That being said, Pagan theology clusters around a few central beliefs that are widely shared. Not every Pagan will embrace all of these beliefs or values, but nearly all will embrace several. The behavior and practices that correlate with these beliefs, as well as the shared identification as "Pagans," are what allows Pagans to maintain a dynamic but cohesive community.

1. *Pantheism, panentheism, or animism.* Pagans experience the divine as immanent in the physical world, although they differ about the details. Pantheists see the world itself as divine ("All is God/dess"); panentheists see the world as immersed in the divine, and vice versa ("God/dess is in all things, and all things are in God/dess"); and animists see all things (or sometimes, all natural things) as having a spirit or soul, but may not believe in a unifying divine force. Some Pagans who technically consider themselves nontheists (in other words, they reject the idea of a literal deity) will still speak in pantheistic terms. For those Pagans, pantheistic language expresses their sense of the

Earth's sacredness and the interdependence of all the beings on and in it.

2. *Polytheism.* Pagans frequently honor multiple deities in their religious practice, both goddesses and gods. Some Pagans are *soft polytheists* and may see the many gods as aspects of one God/dess, or as aspects of a Goddess and a God. Others are *hard polytheists* and understand the gods as individual beings, as separate and unique as human beings are. A number of polytheist positions between—or aside—from these two are also possible, including a nontheist position that acknowledges the many gods as archetypes or personified principles, virtues, or natural forces. Some Pagans honor only feminine deities, while others honor deities of all genders.
3. *Reverence toward nature and the body.* Pagans frequently celebrate natural cycles, especially seasonal changes, and may be passionate environmentalists. Because divinity is believed to be present in the physical world, the body is seen as a sacred part of nature, and sexuality may be considered a special mirror of the divine.
4. *Reference to pre-Christian myths and traditions and/or indigenous traditions.* Most Pagans are converts from Christianity or Judaism, although the number of second- and third-generation Pagans is increasing. Pagans look to pre-Christian religions or indigenous religions that have resisted conversion for what they feel is a more authentic connection to the land, to themselves, or to divinity.
5. *Trust in personal experience as a source of divine knowledge* (sometimes called "gnosis"). With some exceptions, Pagans look to personal experiences as the basis of their beliefs and practices and give experience more authority than texts or received tradition. Intuitive, embodied ways of knowing are emphasized.
6. *Acknowledgment of the principles of magick.* Many Pagans believe that ritual acts performed with intention can alter consciousness and therefore reality. "Magick" is sometimes spelled with a "k" to differentiate it from stage magic and the magic of fantasy novels.[6] Depending on the style and intent of magickal practice, it can have similarities with various activities from other religious traditions, including prayer; healing practices such as laying on of hands; or practices intended to create spiritual changes in physical objects, such as transubstantiation.

7. *Pluralism.* Pagans are often pluralistic both within and between groups. The personal theologies of a group's participants may vary widely so long as the group is able to engage in shared actions. In this way, Paganism shares some qualities with Hinduism, where devotions to the gods may be understood literally by some participants and as philosophical metaphors by others. Pagans often see the traditions of others as equally legitimate as their own. Different spiritual paths exist to respond to differing spiritual needs.[7]

Contemporary Paganism is sometimes called the "Old Religion" because of its attempt to recover the spirit of ancient religious practices. Many Pagans perceive this quality as the ability to cultivate connection between human beings and the land, human beings and divinity, human beings and their ancestors—indeed, between humanity and all of existence. Clearly, however, contemporary people do not and cannot share the worldview of ancient peoples. We have radically different notions about science and the relationship between history and myth than ancient peoples did; we are far less at the mercy of natural forces than they were; and in most cases, we live at a considerable distance from true wilderness. Pagans may draw from what we know of ancient cultures, but this knowledge is being filtered through a modern technological worldview and the lessons of thousands of years of human history. Paganism, then, is also a new religion, one that self-consciously responds to present-day spiritual and environmental crises.

What contemporary Pagan traditions do have in common with their ancient counterparts, however, is the human body. Our circumstances may have changed over the last ten millennia, but our anatomy and physiology have not. Many Pagan traditions center on sensory experience of the physical world and on the ecstasies that can be accessed through dance, prayer, ritual, sex, ordeals, and other physical activities. More than any particular practice or contested historical connection, the continuity of embodied experience links contemporary Pagans with their pre-Christian ancestors.

Although contemporary Paganism draws its body-centered theologies from more than one historical source,[8] many Pagans have been influenced by the originally British tradition of Wicca. In "The Charge of the Goddess," a liturgy edited by Wiccan priestess Doreen Valiente in the 1950s, the Goddess declares that "All acts of love and pleasure are my rituals."[9] This liturgy is widely known among contemporary Pagans, even those who do not practice a Wiccan or Wiccan-derived tradition, and it is often interpreted as affirming all consensual erotic acts. Perhaps because of the influence of this theology, sexual minority status is no bar to leadership in most Pagan communities; in some cases, it is even considered an asset. Although contemporary Paganism has its share of controversies about gender and sexual orientation, many sexual minorities have found it to be more affirming than the religions of their birth.[10]

Pagan Theologies[11]

Theology is a fraught topic for many Pagans. Contemporary Pagan traditions are focused on practice, rather than belief; a group of Pagans gathered for ritual is likely to contain a wide range of theological views, which may not be explicitly discussed. Further, because a significant number of Pagans are converts from creedal monotheisms, many are wary of any theological text or summary of beliefs that could be used as dogmatically. Even the concept of "theology" is often perceived as monolithic and necessarily monotheistic.

However, many Pagans are only dimly aware of liberal Western theologies that focus on issues of justice and on nonexclusive inquiries into the human spiritual condition. From the 1970s through the end of the twentieth century, those in the feminist Goddess movement (including Christians, Jews, and humanists as well as Pagans) produced their own strains of liberal Western theology. Some referred to their writings as "thealogy" to emphasize the distinct nature of the immanent, ever-changing feminine deity (or deities) that they envisioned. Today, many consider feminist theology and thealogy to be an outgrowth of liberation theology, an originally Catholic movement to reinterpret the Christian tradition in relation to socially and economically marginalized peoples. The Latin American liberation theology of the 1950s and 1960s focused primarily on poverty and critiqued society and the Catholic Church through the eyes of the economically disadvantaged. Feminist theology and thealogy began instead with women as its marginalized group and used the lens of gender to critique not just society or the Christian churches, but the whole of Western monotheism. Revaluing women's bodies and affirming women's authority and power was a major focus for much of this work. American black liberation theology, which took African-American experience as its center, was also developed during this time. Later, in the 1990s and beyond, queer Christian and other theologians took a similar approach by critiquing society and dominant religious institutions from the position of sexual minorities. These strains of theology tend to focus strongly on social justice issues rather than on abstract explorations of the nature of deity, and many are *practical theologies*, intended to be applied to public policy and everyday life.

My most immediate theological influences are post-Christian and Pagan feminist theologians who are in conversation with liberation theologies: specifically, Carol Christ, whose foundational essay "Why Women Need the

Goddess" (1978) galvanized the Goddess movement; Starhawk, whose early presentations of feminist witchcraft and its theology are literally practical (her bestselling book *The Spiral Dance* [1979], for example, is a mix of myth, theological exploration, rituals, and spiritual exercises); and Constance Wise, a feminist Wiccan theologian whose *Hidden Circles in the Web* (2008) follows Christ's *She Who Changes* (2004) in its embrace of process theology.[12] Although I do not include process thought as a major focus here, this intellectual movement (sparked by mathematician and philosopher Alfred North Whitehead, then developed into a theology by Charles Hartshorne[13] in the mid- to late-twentieth century) has a great deal to offer developing contemporary Pagan theologies. Many of the basic assumptions of process theology—that divinity is immanent in the world; that all of Being, including its divine aspects, changes constantly; and that human beings are full participants in the ongoing process of divine creation—are widely shared by Pagans.

Like Christ, Starhawk, and Wise, I see myself as engaging in an ongoing Western theological conversation that includes a variety of religious perspectives. Western theology is dominated by Christian voices, and in the United States particularly, the way contemporary Pagans and others hear its originally Greek philosophical terminology is inflected by mainline American Protestant hermeneutics. Yet theology—defined here as an intellectual framework for religious belief and practice—is not a monolithic discipline. Within Christian theology, we find a wide range of social ethics and beliefs about deity, including the body-affirming queer Christian theologies that I engage in this book. Although Pagans, like other Americans, tend to speak of "Judeo-Christian values" or "the Abrahamic religions," such terms threaten to erase the diversity of Christianity, Judaism, and Islam and obscure the commonalities that contemporary Pagans share with some "Abrahamic" practitioners. From the tribal, ecstatic, nature-based practices of Orthodox Jews, to the erotic mysticism of the Sufis, to the veneration of the saints and the Mother of God by Catholics, there are many areas in which Pagans find theological compatibility with those of other faiths.

That "pagan" beliefs and practices appear in religious traditions around the world is one of the points made by Michael York's influential book *Pagan Theology* (2003). As York demonstrates, the beliefs and practices that we have come to associate with indigenous traditions and with contemporary Paganism persist even within traditions that scholars generally think of as monotheistic, exclusivist, and focused on transcendent deity. None of the beliefs and practices of contemporary Paganism are entirely unique (nor, for that matter, are the beliefs and practices of Christianity, Judaism, Islam, Hinduism, Buddhism, etc.). York's goal, however, is not to blur "paganism" (in which he includes contemporary Western Paganism and indigenous religions) with these other world traditions. Instead, he presents its worldview as one of four valid, global religious perspectives: the Abrahamic, the dharmic, the secular, and the pagan. Each position is particularly strong in specific religious traditions: for instance, most Jews, Muslims, and Christians hold Abrahamic attitudes, and most Buddhists and Hindus hold dharmic ones. Yet, as York demonstrates, there are Hindus who hold the pagan position, Jews who hold the secular position, Pagans who hold the dharmic position, and more. For York, the pagan perspective includes multiple deities, the perception that all living things have souls or spirits, the use of images or "idols" in worship, a spirituality that relates to the body or the material, an emphasis on locale and on sacred aspects of geography, and nature worship of some kind.[14] He suggests that these factors appear in so many of the world's religions because they represent human beings' first religious impulses.

I agree with York's implicit observation that Pagan theology does not have to be isolated from the discourse of other faiths in order to be viable. True, there are dangers in using Western theological terminology—most importantly for this book, the possibility that long-standing assumptions around mind-body dualism will contaminate my attempt at a more holistic theology. Critiques of Western dualism, however, are already well underway among progressive Christian, post-Christian, and Jewish theologians. The feminist and queer theologians that I draw on here are actively engaged in alternative approaches to the nature of deity, humanity, and human-divine relationships. Because these interfaith conversations on sexual ethics and the erotic nature of the divine are already in process, this book situates itself primarily in relation to the existing discipline of (liberal) Western theology, and only secondarily in relation to York's sociological treatment of global paganisms.[15]

One potential drawback to my approach is that because of my engagement with so many queer Christian and post-Christian theologians, I find myself inadvertently downplaying the polytheistic aspects of my own theology. Both York and Jordan D. Paper, another religious studies scholar who has turned his hand to polytheistic theology,[16] place this aspect of belief much more at the center of their work than I do here, and I agree that a sense of divinity as multiple is extremely important to developing distinctive Pagan theologies. Particularly to readers who are accustomed to thinking in monotheistic terms, the creation myth that undergirds the theology I present here may seem to present a unitary deity. As I hope will become clear in a careful reading, however, I understand this myth as describing deity as fundamentally multiple—a claim that if there is conscious Being at all, there must necessarily be at least two beings, a Self and an Other. Along with the sacralization of the body and of touch, the emphasis on immanent divinity, and the engagement with Pagan liturgy and myth, I hope readers will keep this implicit polytheism in mind when they consider what in this theology may be uniquely Pagan.

Erotic Theologies

As previously mentioned, positive theologies of the erotic are developing among multiple religious and spiritual communities.[17] Particularly within the past twenty years, writers in the Goddess movement, contemporary Pagan leaders, advocates for the queer and BDSM[18] communities, body-workers, and progressive Christian theologians (including Ellison, quoted above) have all begun to forcefully speak about the sacredness of the body and of touch. Riane Eisler (*Sacred Pleasure*), Janet Hardy and Dossie Easton (*Radical Ecstasy*), Patrick Califia (*Speaking Sex to Power*), Starhawk (*The Spiral Dance*; *The Fifth Sacred Thing*), Deane Juhan (*Touched by the Goddess*), and Carter Heyward (*Touching Our Strength*) have asserted that the erotic is a divinely transformative force, both for personal development and for social change. Although "the erotic" includes sexuality, it is not limited to it; our society's need for and lack of nonsexual connected touch is just as profound as our struggle to enjoy a healthy sexuality. To come into right relationship with the erotic promises not just a lessening of pain, but an opening to profound experiences of joy and connection.

As queer Christian theologian Ellison argues, the erotic is not simply sacred, but also a powerful *moral* force that should inform both individual and social ethics. When the right to pleasure is considered to be a basic human right, acts that do not nurture the body become clear ethical violations in a way American society does not currently acknowledge. For example, though studies have shown that depriving children of affectionate touch stunts them both cognitively and physically (as I will discuss in Chapter 4), rules against touch between public school teachers and children have become increasingly draconian, and some schools have even prohibited hugging and handholding between children. In our efforts to protect children from abusive touch, we have subjected them to the abuse of neglect and encouraged a culture where children and adults alike are agonizingly touch-starved—a state of affairs that encourages illness, intolerance, and violence.

In order to correct this problem, the right to pleasure through consensual touch must be a fundamental value. Making this claim, however, also potentially legitimizes the erotic practices of the cultural and political fringes of mainstream society: those in nontraditional relationships (including same-sex partnerships, open relationships, committed group marriages, etc.); those with sexual tastes often considered deviant (including fetishists, sadomasochists, and others in the kink community); and adolescents, whose erotic behavior is perhaps even more closely scrutinized, criticized, feared, and persecuted than that of these other subcultural groups.

Although this conversation about the erotic as a divine and moral force is happening within various communities, the extent to which it is happening *between* communities is limited. Ellison's book *Erotic Justice* is clearly argued and is specific about how our society's failure to see pleasure as a sacred right supports widespread exploitation and inequality. Published by a small theological press, however, the book primarily draws an audience of Christian theologians and graduate students. Although Christian theologians sometimes read Pagan theologian Starhawk, Goddess movement writer Riane Eisler is often disqualified from the conversation due to her uneven scholarship. Pagans and Goddess feminists may read Starhawk and Eisler, but in turn are frequently dismissive of even progressive Christian theology as being "part of the overculture." While both Pagans and sex-positive activists may read Pagan-influenced BDSM advocates such as Hardy/Easton and Califia, they are largely unaware of spiritual writers on bodywork such as Juhan, who is similarly articulating a moral sense of the erotic without direct exposure to any of these other authors. Many readers and writers who would benefit from these diverse thinkers' works are simply unaware that they exist.

I bring a unique perspective as a massage therapist, Pagan practitioner, and religious studies scholar to this issue. As a bodyworker and a Pagan practitioner, I have personal experience with sharing intimate, nonsexual healing touch with hundreds of people. In my life, the transformative power of the erotic is a reality, not a theory. I have learned a great deal from progressive Christian theology, and I particularly admire the work of Christian sexual ethicists like Ellison because of their clarity and engagement with social issues. Yet because large parts of the Christian tradition have denigrated the body in order (supposedly) to elevate and purify the spirit, progressive Christians often seem forced to write *against* much of their tradition to make their arguments. For me, the task of celebrating the erotic is simpler; as a Pagan practitioner, my sense of the erotic as a moral force is grounded in a cosmology of immanence that draws on widely shared Pagan liturgy and myth (to be explored in Chapter 1).

What to Expect from This Book

Eros and Touch from a Pagan Perspective is a practical theology. It offers readers a cosmology, a system of erotic social and personal ethics, and a basic framework for a mindfulness practice that I've called "sacramental touch." Fundamentally, it is a work of advocacy—not for readers to convert to Paganism, practice deviant sexuality, or seek out professional bodywork necessarily, but rather for the idea that marginalized communities can provide important insight into matters of spirituality and the politics of the body. Liberation theologies propose that we cannot understand our own culture or religion unless we listen to those who are marginalized by that culture. I would argue further that in the realm of bodies, touch, and spirituality, the process of marginalizing sexual minorities in order to defend against their "bad sexuality" produces or reinforces many of our most damaging behaviors. As anthropologist Gayle Rubin writes, "Specific [sexual minority] populations bear the brunt of the current system of erotic power, but their persecution upholds a system that affects everyone."[19] I hope that readers will be able to listen to the voices I bring together here with the knowledge that one need not be queer or Pagan in order to desire—and achieve—greater erotic intimacy with others and with the divine.

This book contains an erotic theology developed from a Pagan perspective. However, because I wish to speak to an audience that is not *solely* Pagan, and because my concerns about touch deprivation and the need to revise our cultural sexual ethics apply to our entire culture, there are important scholarly tasks that I have not undertaken here. Pagan theology is very much a nascent field; although I have described the current state of Pagan theology in detail elsewhere,[20] I am not concerned here with strictly dictating the boundaries of what Pagan theology should be. Further, because this is a work of theology and not religious studies, I have not exhaustively examined the historical context for Pagan values around the body and the erotic. While such a book would no doubt be an excellent contribution to religious studies scholarship, my intent here is to advocate for a religiously-based system of erotic ethics and for policies and behaviors that follow from them. Accordingly, although I will touch on elements of Pagan history and ethnography, they will not be a primary focus.

Readers will also note that in many parts of the book, my approach to the body is clinical and biological, an approach that is unusual in theology and religious studies. Yet I do acknowledge how strongly culture shapes our experiences of our bodies. Much of the theology I draw upon here has been impacted by Judith Butler, who frames gender as a performance through which human bodies are interpreted, and Michel Foucault, who portrays sexual taste, expression, and identity as produced by cultural discourses.[21] In the liberal arts, these two theorists provide a foundation for understanding how culture influences our beliefs about the nature of the body. In including writing on bodywork, with its ultimate grounding in clinical research, I hope to balance the liberal arts' emphasis on social constructivist models by emphasizing the body's biological realities. Robert C. Fuller recently observed that, in religious studies, scholars' overly-strong adherence to cultural constructivist theories has led them to ignore provocative, biologically-based research about cognition, emotion, and sexuality. Fuller rightly argues that the biological sciences provide new information about a wide variety of important religious topics, including the religious functions of pain, the neurochemical basis of spiritual experiences, and the effect of strong emotion on religious territorialism.[22] Clinical research suggests that the flexible framework of culture is supported by the far less malleable ground of our flesh. When I insist that a heightened awareness of the body can provide a basis for ethics, my ideas rest on the insights of biologically-based approaches—an approach that may make some of my colleagues in religious studies and theology uncomfortable. However, clinical approaches and quantitative research methods deserve a place in theology, particularly since I will advocate for concrete changes to our personal and social behavior.

Conceiving an Erotic Ethics

Because of the writing that feminist and queer scholars and sexual minority communities have already produced, I feel able to propose an ethics of touch without including a systematic critique of mainstream American society's sexual mores. This task has been covered by others, and covered thoroughly. In this book, I wish to look to the future. I have lived in communities that are actively raising children with an ethic of sacred eroticism. This book is particularly for those who are exploring the social and community consequences of individual erotic liberation. If you have read this far, I suspect that you already know that the body is sacred; now, I hope to expand your sense of what that knowledge can mean.

Throughout this book, I use the term "erotic" as a category of intimate, embodied, sensual—but not necessarily sexual—contact. Satisfying sex with one's spouse is both broadly erotic and specifically sexual; breastfeeding a baby, on the other hand, is not sexual, but may be intensely erotic. The confusion between the sexual and the erotic is endemic in American culture, and that confusion is one of the reasons most Americans do not get enough affectionate touch. When any sensual, connected touch can potentially be read as sexual, many people refrain from touching for fear of violating personal boundaries. Although one might expect the problem to be less pronounced in more sexually-permissive, progressive communities, in some cases it is actually exaggerated: since progressive communities tend to be more aware of the widespread presence of sexual assault survivors, even appropriate touch may be withheld out of a desire to respect others' healing processes. Distinctions between the erotic and the sexual are often problematically conflated, and so this book will return to the concept of the erotic repeatedly, with special attention to what "eroticism" means in a theological context.

One reader asked me, after reading a passage where I celebrated "the erotic," why I hadn't instead used the word "love." One reason is simply that the phrase "God is love" is so overused that it has become a cliché. It is a bumper sticker slogan that no longer helps people to consider the nature of the divine. Marvin Ellison observes that "love" (as it is most commonly used) also lacks larger connotations of justice. "Love," he says, is "politically ineffectual and reduced to an affective sentiment," nothing more than a sweet, private emotion felt for family members or sex partners.[23] Further, because of the near-ubiquity of the mainstream Christian worldview in the United States, "love" used in a religious context generally means (spiritual, selfless) *agape*, not (desirous, fleshly) *eros*.[24] I propose that divine love is spiritual, intimate, desirous, and carnal, and so I use "the erotic" to highlight its more neglected and maligned qualities.

In writing this theology, I find myself seeking a difficult balance. Many popular sex-positive books and websites disappoint me by glossing over the risks of defying social norms about sexuality and touch. Accordingly, I have tried to be practical and realistic, rather than assuring my readers that all consensual touch is good and a perfect sex life is just one self-help book away. On the other hand, speaking too much about what is practical and realistic, rather than envisioning what would be ideal, makes for an uninspiring theology that is weighed down by the particularities of the moment. As poet Robert Browning puts it, "A man's reach should exceed his grasp, or what's heaven for?"—or as Ralph Waldo Emerson says, more pragmatically, "We aim above the mark to hit the mark."

The principles below form the basis of an erotic ethics, or an ethics of touch. These principles attempt to balance individual rights and needs with collective ones, with the conviction that in a society that cultivates empathy, individual and collective needs can be brought into a productive balance. They are not meant as rigid rules; ethical decisions are complex and informed by individual life experiences, and the decision-maker may be called upon to prioritize conflicting principles. For me, however, these principles flow naturally from the belief that *pleasure is a human birthright.* I offer them as guiding principles for individuals and communities seeking to express that belief in their spiritual practices and relationship structures.

A society based around an erotic ethics:

1. sees *touch as a sacrament*, a way of manifesting divine presence that occurs in a context of self-responsibility, honesty, empathy, and mutuality;

2. is pluralistic, valuing *diversity* of erotic taste and expression as well as of gender expression and relationship structure;
3. prioritizes *pleasure, beauty, and health* in both individual and collective decision-making;
4. maintains community norms that *balance individual freedoms with community well-being* and support economic and social justice work;
5. supports *personal autonomy* through the practices of negotiation, informed consent, and affirmation of individuals' accounts of their experiences.

Queerness and Sexual Minorities

The acronyms used to describe non-heterosexuals and their allies have grown increasingly unwieldy. On several websites, I recently spotted "LGBTQQIAA," which unpacked, means "lesbian, gay, bisexual, transgender, queer, questioning, intersex, asexual, and allies." I have avoided the use of such acronyms because, for political solidarity as well as simplicity, it's important to have pronounceable labels. Based on the usage trends I see in sexual minority communities and also in academia, I've chosen to use the adjective *queer*. Admittedly, "queer" is a still-developing term, and one that gays and lesbians of older generations sometimes still find offensive. Among scholars and younger generations of sexual minorities, however, its quirky, subversive connotations are appealing. "Queer" is now being self-applied—even by some heterosexuals—to indicate a broadly unconventional relationship to gender or sexuality, especially if that unconventionality threatens established social norms. This strategy emphasizes political unity and a common ideological cause around sexual ethics and civil rights. In its contemporary usage, "queer" encompasses unusual sexual orientations that do not necessarily have to do with preferring a given gender or a particular set of taboo sexual practices. The term indicates a more general fluidity and unconventionality in gender presentation or identity, sexual orientation, and/or sexual role. (As I will argue later, however, the entire notion of "conventional" sexuality is a shaky one. Human sexuality is incredibly diverse, and the tastes that are clustered together under terms like "heterosexual" or "vanilla" are not as uniform as they seem.)

This work is also situated in the context of queer studies and queer theology because of the limitations I see in feminist theory and theology. I consider myself to be a feminist, and many of the issues I have with second wave feminism (the feminist writing and activism of the 1970s and 1980s) have been addressed by third wave feminism (in the 1990s and early twenty-first century). Second wave feminist writing often treated gender as a binary category: society was seen as dividing human beings into "men" and "women" and then setting up a hierarchy between them with women on the bottom. This hierarchy, then, had to be overturned and rebalanced so that women could have the same economic, political, and social freedoms that men do. To the extent that some societies (and some parts of US society) still sharply divide human beings into two genders, much second wave feminist analysis is still valid. Third wave feminism, however, sees the binary as the problem. Rigid gender roles limit and damage all people; the diversity of gender expressions and identities that human beings are capable of do not fit well into a two-category system. The problem is not simply securing equal rights for women, but rather creating a society in which individuals' life choices are not prescribed or limited based on their genitals. The fluid model of gender that both third wave feminist writers and queer studies writers employ fits my experience of contemporary, urban American society far better than the binary model of the second wave.

Queer and later feminist theory has also become increasingly sophisticated on the topic of intersectionality, a method that highlights the diversity of social roles and modalities that define groups' and individuals' positions in society.[25] Intersectional approaches help to avoid the simplistic applications of gender binaries that can be found in some academic work even well into the 1990s. For instance, in researching my dissertation project, I came across a book that uses a gender-binary feminist theory to approach the issues of religion and sexuality in American literature. The author divides American novels into two categories: "androcentric" (male-centered) and "gynocentric" (female-centered). Here, sexuality is framed in the clichéd and essentialist terms of the gender wars: women's sexuality is nurturing, egalitarian, and enlightened, while men's sexuality is domineering, hierarchical, and oppressive. The author is careful to point out that not all novels written by men are "androcentric" according to her model. Still, the book's binary model implies that women are universally victims, and also that they also possess greater erotic complexity than men. "Gynocentric" is clearly a superior point of view in this book, even if "women" are not necessarily a superior group.[26] This simplistic model of gender, sexuality, and desire obscures the complex movement of power and desire among individuals in contemporary Western society. "Men" and "women" are still significant social categories, yes; but differences in ethnicity, economic class, education level, sexual orientation, and gender expression are also important factors and change the dynamics of power. This awareness is central to the method of intersectionality. Among a diverse group of people in a workplace, for example, assuming that power lies primarily with men is likely to be incorrect, as multiple hierarchies of privilege are operating at once, including race, class, gender, nationality, educational level, sexual orientation, age, and more. Among a white, highly educated lesbian, a gay Latino man, and a working-class Russian man who speaks English as a second language, who will be in charge? Further, who will desire whom, and in what way? For both power and desire, so much depends on context. Third wave feminism, accordingly, has attempted to take race, class, and sexual orientation more completely into account, while queer studies provides more flexible models for gender and a rich vocabulary for the relationships between gender and sexuality.

The word *queer*, however, does have some problems, and for that reason I also use *sexual minority* as a term that is broader and more neutral in tone. I mentioned already that some gay and lesbian people still hear "queer" as a slur. These individuals are often less countercultural than their queer-identified counterparts; they simply want same-sex partnership to be a socially acceptable relationship choice. Non-queer-identified gay and lesbian couples often seek to replicate traditional American family structures with a same-sex partner. Queer-identified individuals are more likely to actively cultivate alternative family structures, including nonromantic partnerships (such as a pair of sisters raising children together), multiple-partner romantic relationships, and single people in collective households. According to some queer activists, institutionalized same-sex marriage may reinforce the same problematic social values as heterosexual marriage: unexamined monogamy and the isolated nuclear family.[27]

The result is the perception that one can be gay or lesbian without being queer. As the terms are being defined, queerness often involves an inherent opposition to the social and political values of mainstream society, but gayness does not. The fact that social attitudes about homosexuality have shifted profoundly is cause for celebration. In some places in the United States, a person can be openly homosexual and attend a mainstream church, join the PTA, run a successful business, or participate in the local Republican caucus. But the increasing prominence of the word *queer* is an indication that for many individuals, a minority sexual or gender identity is still synonymous with countercultural (or at least progressive) politics. The continuing heated debate over same-sex marriage legislation also demonstrates that in many parts of the United States, homosexuality is far from socially acceptable. I employ the term *sexual minority*, therefore, in order to include gay and lesbian people who (for whatever reason) do not identify with the word "queer."

Sexual minority also includes those individuals whose gender identity and sexual orientation are conventional, but whose sexual practices are marginalized nonetheless—those who participate in the leather, kink, BDSM, and fetish communities, or who engage in such practices privately with spouses or lovers. Individuals choosing celibacy for religious or personal reasons also fall into this category; though they may not see themselves as "queer," their sexual choices may nevertheless disturb or upset their families and friends. Those whose sexual desires or practices differ from the mainstream are often discriminated against, pathologized, and pressured to conform, with the result that many sexual minorities must protect their sexuality as a dangerous secret.

It is also possible for sexual minorities to be produced by a rigid social context: for example, if an individual's beliefs or attitudes about the meaning of sexuality, sex, or gender significantly differ from those of his community. In the autobiographical graphic novel *Blankets*, for example, the teenaged Craig Thompson is portrayed as sweet, sensitive, and artistic—and therefore, automatically suspected of being gay by the aggressively masculine football players in his small Midwestern town. His developing sense of his own sexuality as spiritual is utterly at odds with the religious culture of his family and community, where sexuality is considered either a dirty, guilty pleasure or an outright sin. Craig experiences regular harassment and accusations of queerness because of his subtle differences from his peers.[28]

As in this example, even while being heterosexual, an isolated individual in a culturally homogeneous community may experience a similar alienation, conflict, and sense of being closeted that LGBT persons do. This situation is largely due to our culture's lack of a pluralistic erotic ethics. Against the mainstream assumption that sexual variation is culturally destabilizing or presents a moral threat, I will argue that most variation in sexual and erotic tastes is benign, and that a healthy society should protect adults' right to engage in consensual, pleasurable acts.

Paganism and Sexual Minorities

Although contemporary Paganism is not without its own homegrown strains of homo- and transphobic theology, the movement as a whole has been welcoming to sexual minorities. The notions that the body and sexuality are "sinful" or that the only divinely approved sexuality is heterosexual have driven many people out of mainstream religious institutions and left them seeking a spiritual home.[29] Paganism's openness to a variety of forms of sexuality has attracted significant numbers of LGBT people. An extensive survey published in 2003 found that 28.3 percent of American Pagans identified as gay, lesbian, or bisexual.[30] Current statistics estimate that 4–10 percent of the general US population identifies in this way,[31] so this concentration of sexual minorities in the Pagan community is significant.

This book gives particular attention to three communities of sexual minorities: polyamorists, BDSM practitioners, and transgender people.

I have done this partially because these groups have developed a strong internal sense of identity: there are many organized and semi-organized groups for these communities, and all have articulate writers advocating for their concerns. Nevertheless, they are much more socially and politically vulnerable than the gay, lesbian, and bisexual communities, which have made great strides forward in the last twenty years. The more compelling reason for my choices, however, is that these groups have a special status within contemporary Paganism. Although all three remain controversial among Pagans, they also enjoy a greater level of either passive tolerance or active acceptance than in society as a whole, and this has allowed members of all three communities to engage contemporary Pagan theology to describe their sexual or gender identities as sacred. I bow to popular practice here in including transgender as a "sexual" minority, although in reality transgender people do not engage in any specific sexual practice, nor hold any particular sexual orientation. My experience with these sexual minority communities, however, is that transgender and sexual orientation issues are seen as tightly related.

Transgender

"Transgender," like "queer," is a mutable term for individuals who deviate from conventional gender roles. I use it primarily to refer to two groups. The first group includes individuals who are genetically male and have altered (or are altering) their appearance in order to present and be accepted as women (male to female, or MTF), as well as individuals who are genetically female and have altered (or are altering) their bodies to present and be accepted as men (female to male, or FTM). Many of those who choose to transition feel that they were born into a body of the wrong sex, and hormonal treatments, surgical sex reassignment, or a combination of the two frequently resolve their gender dysphoria. The second group consists of those who consider themselves "third gendered"—they do not feel comfortable as either "men" or "women" and may present as androgynous or with strong elements of both traditional masculinity and femininity. Some third-gendered individuals choose to alter their bodies, but others do not. As surgical procedures for FTM transition remain inadequate, and procedures for both MTF and FTM transitions are expensive and may impair sexual functioning, not every individual who wishes to transition opts for surgery.

"Transgender" can also be used more broadly to include transvestites (cross-dressers) and drag kings and queens (individuals who occasionally cross-dress in gender-exaggerated costumes, but who present more conventionally in everyday life). The related term "genderqueer" can also be used for individuals whose standard gender presentation deviates from conventional gender roles. For instance, a gay man who is usually perceived as male, but who identifies as "femme" and is sometimes taken to be female by strangers, might consider himself to be genderqueer.[32]

Despite the fact that many Pagan communities are welcoming toward transgender people, transgender issues have been controversial in Paganism because of the strong influence of British Wiccan theology on the Pagan movement as a whole. British Traditional Wicca focuses on fertility and uses heterosexual sex (the union of the Goddess and the God) as a symbolic representation for the interaction of primal creative forces.[33] Transgender people do not fit into binary gender roles, which has left some Pagans at a loss for how to include them in rituals that call for a priestess and a priest. Some Wiccan covens understand masculinity and femininity as forces within every individual, and so allow individuals of any gender to take on either the Goddess or God role. Others welcome fully transitioned transgender individuals, but are uncomfortable with those who don't "choose a side," so to speak.

Pagan gatherings have also sparked controversy when transwomen and transmen have been excluded from rituals exploring "Women's Mysteries" or "Men's Mysteries." Particularly among female Pagans who worship only a Goddess, some believe that the experience of growing up with a womb and vagina has fundamental and wide-ranging effects on women's perspectives and spirituality, and that these are experiences that transwomen cannot share. In contrast, others feel that experiencing the menstrual cycle is a non-essential aspect of being a woman. Focusing on anatomy as the only source of gender mysteries can exclude women who have never menstruated or have had hysterectomies, and women who do not want or cannot have children may feel marginalized in groups that see childbirth as a central women's mystery. As in the wider culture, there is disagreement about whether one is born a woman, or whether one is made into one by a combination of physiological and social processes. From the latter point of view, women's experiences of womanhood are inherently diverse, and so the womanhood of transwomen can be understood as one valid form among many.[34]

Some Pagans explicitly affirm the fluidity of gender in their theology and have created new myths and rituals that do not depend on a gender binary. For instance, in the creation myth that I analyze in Chapter 1, the Goddess is not truly female, but contains the potential for all genders (and is perhaps pre-gender). Other resources for constructing innovative queer and transgender Pagan practice include Raven Kaldera's *Hermaphrodeities*, a book of myths, rituals, and reflections on transgender and genderqueer gods from pantheons around the world, and P. Sufenas Virius Lupus' *All-Soul,*

All-Body, All-Love, All-Power: A Transmythology, an epic poem that draws on the mythology of the ancient Hellenistic world to describe the birth of a transgender tetrad of deities.[35]

Polyamory

Polyamory is a neologism for the practice of ethical nonmonogamy (a situation where all participants are fully informed about and enthusiastically consenting to a multiple-partner sexual relationship). As a religious phenomenon, polyamory has precedents in several American new religious movements: in the nineteenth century, Mormon communities practiced polygamy (and some groups continue it in private today), while the Christian Oneida commune experimented with "plural marriage," a system in which monogamous sexual commitments were forbidden as potential distractions from devotion to God.[36] As practiced by contemporary Pagans and other largely left-leaning alternative people, polyamory is gender-egalitarian and can include a variety of relationship types. Polyamorists may engage in "polyfidelity," a kind of group marriage where partners do not engage in sex outside of the group. They may divide relationships into "primary" or "secondary," as when a woman with a committed partner (her primary) also has a more casual girlfriend or boyfriend (her secondary). Or, they may have entirely open relationships, where individuals may be casually or seriously involved with a number of different sexual and romantic partners. Polyamorists emphasize honesty, informed consent, and safer sex practices. Unlike the swinging community, where most participants are heterosexual couples, polyamorists are diverse in their sexual orientations and relationship status. Although some are interested in having sex for fun rather than as part of a developing relationship, polyamorists tend to emphasize the pleasure and thrill of becoming emotionally attached and falling in love. Communities of polyamorists in their thirties and forties are frequently family-oriented, and long-term partners may trade babysitting duties so that their significant others can go out on dates.[37]

Many polyamorists were first introduced to the idea of ethical nonmonogamy by Robert A. Heinlein's 1961 science fiction novel, *Stranger in a Strange Land*. This book was also a direct influence on polyamory's development in contemporary Paganism. In 1962, Oberon Zell-Ravenheart (then Tim Zell) and Lance Christie founded the Church of All Worlds (CAW), a Pagan group modeled on the organization of the same name that Heinlein describes. In CAW, they embraced the radical honesty, communal living, and ethical nonmonogamy that they admired in *Stranger*. As in *Stranger*, the small groups making up CAW referred to themselves as Nests and practiced water-sharing as an individual and community ritual of bonding. In the 1970s, CAW produced the influential Pagan magazine *Green Egg*, which showcased some of the important voices of the early American Pagan movement and served as a networking tool for like-minded spiritual seekers.[38]

After two decades of sustained (if imperfect) communal and non-monogamous living, Zell's partner Morning Glory Zell coined the term "poly-amorous" to describe the practice of consensually nonmonogamous romantic relationships.[39] The term "polyamory" came into wider subcultural use when the Usenet group alt.polyamory was founded by Jennifer Wesp in 1992.[40] Since the early 1990s, the term and the practice have attracted growing numbers of adherents—many Pagan, though some not. Dossie Easton (who identifies as Pagan) and Janet W. Hardy's 1997 polyamory how-to *The Ethical Slut* sold over 50,000 copies by 2006,[41] and it has now been released in an updated second edition.[42] Amazon.com showcases a steadily growing selection of titles on polyamory for Pagans, heterosexuals, lesbians, and other special interest groups. Although sociologists Helen A. Berger, Evan A. Leach, and Leigh S. Shaffer report in their 2003 survey that only 0.4 percent of American Pagans are living in formal group marriages, support for group sexual relationships in Pagan communities is widespread; 51 percent of Pagans surveyed expressed support for the legalization of polygamy, while 24 percent held no opinion (only 25 percent were opposed).[43] Notably, in her 2004 book *Rites of Pleasure: Sexuality in Wicca and NeoPaganism*, Jennifer Hunter finds herself not in the position of trying to justify polyamory as a Pagan practice, but of arguing for the equal legitimacy of *monogamy*.[44] Monogamous Pagans sometimes feel marginalized by their perceptions of a polyamorous majority; in the mid-2000s, one large Pagan festival in Central Texas featured an affinity group for monogamous Pagans self-deprecatingly named "Dry White Toast."

Pagan polyamorists often see polyamory as a natural consequence of their belief that all acts of love and pleasure are the rituals of the Goddess (as stated in the well-known Wiccan liturgy by Doreen Valiente, "The Charge of the Goddess"[45]). Most polyamorists freely admit that the communication, self-awareness, management of insecurity and jealousy, and sheer difficulty of social scheduling make polyamory very hard work. They see this work as worthwhile, however, and sometimes even as part of a religious calling: polyamory allows them to fully express sacred desires, rather than suppressing or even denying them to protect a monogamous relationship. Additionally, many value the densely interconnected community that polyamory can create and speak of their loved ones as their "chosen family" or "tribe."[46]

BDSM

BDSM is an acronym for "bondage and discipline/dominance and submission/sadomasochism." The BDSM community is sometimes also referred to as the "kink" or "leather" community and includes those who enjoy a variety of intense erotic practices, such as physical restraint, spanking, flogging, or role-playing where participants take on the roles of "top and bottom" or "dominant and submissive." Dominance/submission tends to emphasize the role-playing element, while sadomasochism focuses more on physical sensation.

In *Rites of Pleasure*, Jennifer Hunter provides a list of key points for understanding BDSM both in general and in a Pagan context. She quotes Dossie Easton on the use of the word "play" by BDSM practitioners to describe their activities: "There's play like little children playing, there's play like I'm not real serious, then there's play like drama and theater, and there's play like psychodrama. And there's play like dancing with the universe."[47] BDSM play can have elements of all of these: it can be silly and fun; dramatic, with costumes and props; deeply psychologically revealing; or spiritually transformative. Hunter also emphasizes that healthy, consensual BDSM has nothing to do with abuse. Participants can end the activity at any time with the use of a pre-defined safeword. Although pain may be involved, causing pain is not the point of BDSM; BDSM is fundamentally about pleasure, especially the pleasure of two equals exchanging power. Hunter also observes that BDSM practices are much more common than most people think: according to the 1990 Kinsey Report, BDSM activities are practiced by 5 to 10 percent of the US population. The report notes that "most incidents [are] either mild or staged activities involving no real pain or violence."[48]

As with the issue of polyamory, many Pagans hold that people have a fundamental right to pleasure. Consensual pleasure between adults needs no justification. Pagans tend to be suspicious of mainstream culture's negative attitudes toward sex and the erotic, as clearly harmless forms of eroticism such as oral sex and homosexuality have been historically demonized. Many are accepting of mild forms of BDSM, and some traditions even use the light application of a soft flogger as a trance induction technique. But when it comes to more extreme forms of BDSM—often called "edge play"—Pagans tend to be almost as skeptical as mainstream society. Whipping, putting hooks or needles through the flesh, master/slave scenes, and heavy impact play are particularly criticized by some feminist Pagans as reinforcing the patterns of an abusive patriarchal society.

Pagan writers on BDSM have responded to these criticisms by focusing on the essentially egalitarian nature of BDSM relationships (contracts made between two equals, each of whom can end the contract), drawing parallels with traditional shamanic practices, and emphasizing the healing and empowering experiences that BDSM offers to both bottoms and tops. Dossie Easton, for instance, works professionally as a licensed counselor, and in her books with Janet W. Hardy, the two writers focus on BDSM as a method to find or reclaim one's own power. Bottoms can demonstrate their own strength by undergoing ordeals or facing deep-seated fears; tops can experience the expansive combination of nurturing and discipline that it takes to skillfully facilitate a bottom's experience. Other motivations for BDSM practice often include emotional catharsis, achieving altered states of consciousness, or sheer enjoyment of physical intensity. In *Radical Ecstasy*, Easton and Hardy particularly compare BDSM to consciousness-altering tantric practices, through which participants have profound experiences of divine presence.[49] Similarly, in *Dark Moon Rising*, Raven Kaldera frames BDSM as a way of using the body to gain greater access to spirit. Traditional forms of shamanic BDSM practices include Catholic practices of flagellation, as well as the Lakota Sun Dance and Hindu ball dances (both of which involve inducing altered states through experiencing intense, extended, but non-life-threatening pain).[50] Kaldera, who has published widely on alternative sexuality and gender expression from a Pagan perspective, emphasizes the role of the top as a technically skilled, spiritually knowledgeable facilitator who helps participants engage in such practices as safely as possible. He also speaks of the role of BDSM in initiation, where a psychological or physical ordeal is undergone in order to confront one's own mortality, overcome a fear, or otherwise gain wisdom about the self. Kaldera's view is supported by neuropsychological research into the religious uses of pain, where pain applied in a ritual context has been shown to reliably mediate changes in sense of self, motivations, and priorities.[51]

Pagan and radical sex activist Patrick Califia has a similarly religious approach to BDSM (or, to use the term he prefers, sadomasochism). For Califia, sadomasochism expresses the belief that "Pleasure is a blessing [and] delight is our sacred birthright."[52] Recently, he has presented his religious perspective as a member of the BDSM community and a Pagan at two meetings of the American Academy of Religion.[53] Califia and Drew Campbell's collection *Bitch Goddess: The Spiritual Path of the Dominant Woman* and Lee Harrington's more recent book *Sacred Kink* both discuss the use of BDSM for Pagan devotional purposes.[54] In the practices these writers describe, participants undergo ritual ordeals as a way to gain a deeper understanding of key religious myths, such as the descent of Inanna. Ordeals can also serve as gifts to divinity; for example, a Pagan might be tattooed in sacred space with a design that honors a particular deity. The spiritual practice of BDSM is not limited to Pagans, however. Mark Thompson's collection *Leatherfolk* includes essays from a variety of spiritual perspectives, including pieces by a Christian minister, three Pagans, and several practitioners of Eastern-influenced paths.[55] Further, the practice of BDSM may become spiritual, even if it was not originally framed in those terms: several of the above writers remark that intense BDSM activities can lead to a deeper interest in religious practice, even if all practitioners were initially looking for was exciting sex. (For further discussion of BDSM in a religious context, see Chapter 3.)

Even if BDSM remains controversial in the Pagan community, it is still far more acceptable there than in any other major contemporary religion. Pagans' reverence for nature and longing for greater affinity with the animal world allows them to understand themselves as animals with primal drives. BDSM provides a structured container in which some of those drives can be explored.

Consent

In this book, and in the theological traditions with which I'm engaging, the erotic is the mutual, pleasurable, fully embodied movement of energy between beings. A long, affectionate embrace between friends or lovers is an easy example of simple eroticism, as is petting a cat. One might also experience the erotic without a human or animal companion, however. Lying peacefully in warm sunlight or skinny dipping at night can also be erotic—the partner or partners in the exchange are elements of the natural world. Nor does physical touch necessarily have to be involved if the encounter still engages the body. Catching the eye of a stranger at a party and feeling a rush of connection and attraction is an intensely erotic experience that causes a whole-body reaction, yet involves no physical contact. When we are swept up by the erotic, denying, forgetting, or neglecting the body becomes impossible; its centrality to our existence cannot be ignored.

I am explicitly excluding nonconsensual acts between human beings from the category of the erotic. The theological traditions that I draw on, beginning with C.G. Jung, all understand eros specifically as a relational principle that binds individuals and groups together. As a Pagan, I do believe that there is life energy in the non-human world that binds together predator and prey, the eater and the eaten. The self-awareness and possibility of moral choice that separates human beings from animals, however, forces us to participate differently in that dance. Because human beings are still a type of animal, we retain something of the merciless savagery of predators. Yet human beings also possess empathy, a trait that (in the absence of major trauma or psychological disorders) is easy to cultivate and requires heavy conditioning to stamp out. As recent studies on the psychology of killing suggest, even soldiers who have been trained to kill frequently fire their guns without aiming, so reluctant are they to harm the enemy.[56] Studies of post-traumatic stress disorder also show that the soldiers who are most traumatized by combat and have the most severe and lingering symptoms are those who witnessed or engaged in wartime atrocities.[57]

For most human beings, to knowingly violate other people requires objectifying and dehumanizing them. Although humans—particularly when damaged by repeated trauma—may experience pleasure in violence, such behavior is associated with illness and a lack of functionality in other areas of life. Even in human relationships where violence does seem to cause a kind of bonding, as when a battered partner refuses to leave his abusive spouse, the isolation, anxiety, depression, and suicidal impulses that abuse victims often experience demonstrate the relationship's fundamentally destructive effects.[58] The erotic does not flow when pleasure is one-sided; the pleasure must be mutual and encourage a deeper, more life-affirming connection. Some Native peoples believe that prey animals ultimately surrender to the predator, and that the relationship between predator and prey helps to keep nature in balance, for the ultimate good of all species. Yet when a human being violates, abuses, or kills another, no hungry mouths are fed; there is no balance that is struck, no culling of the herd that allows the whole to flourish. For human beings, nonconsensual contact results in a constriction of life force that is antithetical to the erotic as I discuss it here.

The issue of defining eroticism is muddied by our culture's lack of clarity about what constitutes consent. In the recent collection on rape prevention and female sexuality *Yes Means Yes!*, the point is repeatedly made that consent is not the absence of a "no"; it is the presence of a "yes." This point was driven home to me a few years ago when I was co-teaching a workshop, one of the themes of which was self-responsibility. As we approached the end of the day, it became clear to my co-facilitator and me that we would need fifteen extra minutes for the day's last exercise. On a break, we called the group together and asked if it would work for everyone to extend the class. There was a general nodding of heads, and we allowed the class to run overtime. The next day, a student approached me and told me how angry she was that the class had run late and how greatly it had inconvenienced her. I was momentarily taken aback, thinking that we had somehow not noticed her missing from the room when we checked in. She replied that she had been in the room, but hadn't felt able to go against the rest of the group in expressing her need. (Later, she apologized for essentially expecting the facilitators to read her mind.)

This incident illustrates the kinds of common cultural dynamics that lead basically kind, caring people into situations that feel consensual to one party and nonconsensual to another. When we asked to extend the class, we took a lack of "no" as consent. Since all the students were adults and free to leave at any time, and the facilitators were in no strong position of authority, it was probably reasonable of us to expect the students to advocate for their own needs. But when it comes to obtaining consent to touch, the reality of our culture is that damaging assumptions are widespread and the risk of harm is high. Sexual assault is common, particularly the kind of nonmalicious sexual violation where the person initiating sexual touch believes the touch is desired, while the recipient is actually uncomfortable, embarrassed, or frightened. Sadly, men are still often socialized to believe that a "nice girl" always refuses sex, and that it is a man's role to convince or persuade her. In this context, a weak or uncertain refusal may be taken as a demonstration that the girl is not "easy"—often considered a valuable trait—and silence and passivity during sex are seen as normal. In a culture where many people find it difficult or embarrassing to discuss sex at all, the initiators of sexual touch rarely seek a clear "yes," while the recipients are unprepared to say either "yes" or "no." To further confuse the issue, people commonly engage in sex they do not want for a variety of emotional and social reasons (for example, to preserve a relationship; to prevent a fight; to fit in with a peer group). As Margaret Cho relates in her foreword to *Yes Means Yes!*, one can even get into the habit of *initiating* unwanted sex—and, over time, become increasingly unable to tell when sex is actually desired.[59]

A friend tells me that if he's going to have sex, nothing less than enthusiastic participation from his partner will do; if he can't get a clear affirmation, particularly from a first-time partner, he ends the encounter. I've come to see this principle as essential for healthy sexual relationships, as well as a foundational for healthy nonsexual touch: the erotic requires *participation.* In a culture where the recipients of touch are expected to be passive, many people become so accustomed to accepting unwanted touch that neither initiator nor recipient can easily tell when touch is actually desired, as opposed to simply endured. Transgender writer Cedar/Hazel Troost tells the story of what happened when s/he asked hir friends to get permission before touching or hugging hir: many of hir friends resisted on the basis that friends are entitled to touch one another at any time.[60] Some, however, cooperated, and Troost reports that having the opportunity to refuse touch—but even more so, having the opportunity to say "yes" to it—had the effect of clarifying desires that an abuse history had made murky. In the course of insisting on explicit verbal consent, s/he realized that s/he did indeed want a great deal of touch, perhaps even more than s/he had previously been receiving. Having the opportunity to say "yes" put hir in touch with hir desire.

The sense of entitlement that Troost's friends had about hir body was distressing to discover, yet unfortunately typical. The unexamined belief that an existing relationship—friendly, familial, or romantic—provides an automatic and irrevocable consent to touch is common, and it creates the conditions for confused and violated boundaries. Exploitation and abuse are also more easily justified when others' fundamental right to control their own bodies is not acknowledged. In contrast, the practice of obtaining explicit verbal consent for touch helps to ensure that the contact is desired and enjoyable for both the initiator and receiver. If consent is never assumed except by prior agreement and can be withdrawn at any time without automatic risk to the relationship, it becomes much more likely that contact will be whole-hearted and connected—essential conditions for an experience of the erotic.

Those who are skeptical that BDSM can be a healthy and fulfilling sexual practice are frequently concerned about the validity of participants' consent. It is true that BDSM can involve stresses on or injuries to the body. For some, consenting to be harmed is automatically a sign of mental illness, and mental illness negates the validity of consent; the person who desires harm is not in his right mind, and for the sake of his well-being, some authority (whether family, the medical establishment, or the law) should intervene in his choices. BDSM practitioners, however, believe that their consenting to intense physical practices is an expression of a moral right to autonomy. In examining the ethics of consent in BDSM, ethicist Morten Ebbe Juul Nielsen affirms autonomy as a defensible moral good. Situating his argument within an ethical tradition that values individual freedom and responsibility, Nielsen argues that a significant level of personal autonomy is necessary for human flourishing and well-being.[61] Consent, then, draws its moral force from this basic human right.

Yet as Nielsen shows, if consent is based on human beings' right to personal autonomy, consent cannot be given for acts that destroy that autonomy.[62] Without autonomy, people cannot take personal responsibility for their choices and so have no capacity to consent to any action or agreement. Nor does the act of having given consent entirely transfer responsibility to another party. If a person consents to have sex with a partner, he is responsible for choosing a partner who is appropriately morally responsible, and for withdrawing his consent if his judgment turns out to have been flawed. The BDSM concept of the safeword, which can be invoked at any time, emphasizes that even when a participant appears to have given his power away, he retains the right and the responsibility to end any encounter that doesn't fit his desires.[63]

Nielsen points out, however, that consent can be validly used to compromise autonomy on a temporary basis. For example, it is commonplace for people to consent to surgery that involves full anesthesia, even though they are not autonomous while unconscious.[64] Such procedures are often undertaken to save lives, in which case the temporary state of being anesthetized and then in recovery preserves the individual's autonomy in the long run. However, it is also widely considered acceptable to undergo surgery to relieve pain or increase performance, or simply for cosmetic reasons. Here, the right to govern one's own body justifies the temporary loss of agency caused by the medical procedure. Through this example, Nielsen illustrates the fact that in Western society, individuals already consent to compromises of their autonomy on a routine basis without completely undermining the basis on which consent is given. If the compromise of autonomy is temporary and contributes to greater autonomy in the future, it is morally legitimate for an individual to consent to a loss of control.[65]

Nielsen uses this ethical framework to argue for the moral legitimacy of consent in BDSM. One can consent to power exchange (temporary loss of control) and to intense physical experiences that might cause injury because the individual has the right to govern her own body. Additionally, because BDSM practitioners experience these activities as producing self-knowledge and pleasure, they potentially increase individual autonomy in the long run. In addition to the reinforcement of autonomy, Nielsen also specifies additional factors for ensuring valid consent: first, that it must be informed consent, with both parties fully understanding the activities they are agreeing to perform and their risks; second, that consent be intentional and wholehearted, not given as a bargaining chip to get something unrelated out of the relationship; third, that it be voluntary, not compromised by coercion (threats of harm); and fourth, that the person giving consent is competent (not intoxicated, under extreme emotional duress, immature, etc).[66] Although all of these factors—especially competence—have a degree of subjectivity, they provide a framework to evaluate whether an individual's autonomy has been significantly undermined.[67]

Additionally, Nielsen discusses exploitation and manipulation as relationship factors that do not undermine the ability to consent but which are morally wrong for different reasons.[68] Nielsen defines exploitation as a situation where one party takes unfair advantage of another's vulnerability or natural proclivities.[69] For example, imagine a situation in which a woman leaves her husband, then agrees to return to the relationship only if the husband cedes total control of the family's finances to her. The husband is not being coerced with threats of force, but his desire for his wife may lead him to allow himself to be exploited in this way. Manipulation similarly does not undermine autonomy, but involves deception, without which consent might not be given.[70] A woman who has sex with a man she believes is a millionaire, but who is actually a used car salesman, has been manipulated and wronged, but her consent to sex is not invalid; she has been deceived, but not raped.

BDSM relationships are subject to the same ethical rules as other kinds of sexual relationships: there must be an absence of exploitation or manipulation, valid consent must be given, and both individuals must retain the ability to opt out (as Nielsen puts it, each must have a "real opportunity to exercise a robust exit right").[71] If a BDSM relationship meets these conditions, then the right of the individual to autonomy (and by extension, to self-determination and individual flourishing) can justify the relationship's risks.

Nor is this the only area of life where risky but regulated behavior is socially condoned. Within the United States and internationally, for example, there is widespread debate about the relative benefit and harm of various drug laws. The Netherlands has firmly come down on the side that, at least when it comes to marijuana and naturally occurring psychedelic substances, permissive laws protect both society and the individual better than draconian ones. The United States has its own complex history with drug legislation, with the passage of the Twenty-First Amendment attesting that the repercussions of alcohol prohibition are more harmful to society than alcohol use itself. Although alcohol use has negative social and physical effects, those negative elements are balanced by its ritual role in some religions and the enjoyment that many people experience from responsibly consuming it. Further, legally prohibiting its production and use infringes heavily on the autonomy of individuals and tends to create a dangerous black market. Though alcohol can hardly be said to be harmless, and it is implicated in a wide range of illness and social dysfunctions, it is also clearly possible for people to use alcohol in a healthy and responsible fashion. Accordingly, legal adults are permitted to produce, distribute, and consume alcohol within certain restrictions—restrictions ideally meant to minimize harm.

In American culture, it is widely accepted that drinking alcohol is a somewhat risky, but pleasurable activity that can be engaged in responsibly. The right of adults to use alcohol within the limits of the law is assured. Although studies demonstrate that frequent alcohol use is harmful to the user's health, we do not question the validity of an adult's consent when she buys herself a drink. It is certainly possible to abuse BDSM in similar ways to alcohol—there is a possibility of addiction and the aggravation of psychological disorders. Yet with both activities, there are potential benefits to both individuals and communities. Alcohol can be consumed joyfully, in a toast at a wedding that loosens inhibitions and encourages the guests onto the dance floor, or shared between lovers as part of a night of pleasure; BDSM can create experiences of intense connection and intimacy, or provide opportunities for practitioners to explore their power or increase their self-confidence. The existence of alcoholism and abusive relationships does not negate the potential for either alcohol use or BDSM practice to be life-affirming.

The assertion that a pleasurable activity is inherently harmful requires evidence that its risks outweigh its benefits; without evidence, this assertion is merely prejudice. In the absence of reliable clinical and statistical studies of BDSM practitioners, most criticism of BDSM has relied on "common sense"—if it looks like violence to outsiders, it must have the effects of violence on practitioners. As the BDSM community has become a more vocal advocate for itself, however, psychologists and researchers have begun to take a greater interest in the community and to conduct formal studies. Gabriele Hoff and Richard Sprott's 2009 study of BDSM practitioners who seek therapy argued that, although BDSM sexuality can be expressed non-pathologically, BDSM practitioners' access to effective mental health services was limited by the prejudices of therapists, many of whom assumed that BDSM sexuality was an illness.[72] This prejudice among therapists lingers despite shifts in the standard diagnostic guidelines for mental disorders. In the most recent editions of the *Diagnostic and Statistical Manual of Mental Disorders* (the DSM-IV-R and the forthcoming DSM-V), several consensual sexual practices that were once considered "disordered" (such as sexual masochism and sadism) are now listed as pathologies only if they cause "clinically significant distress or impairment in social, occupational, or other important areas of functioning" or involve a violation of consent.[73] Additionally, a 2001–2002 national phone survey of Australians on sexuality produced statistical data suggesting that BDSM sexuality does not correlate with mental illness. The 2.2 percent of Australian men who admitted to enjoying BDSM activities scored significantly higher on a measure of psychological well-being than those who did not report practicing BDSM. The 1.3 percent of BDSM-practicing women reported no greater incidence of depression, anxiety, or sexual difficulties than the general population, and were no more likely to have been coerced into sexual activity. BDSM activity was higher among gay, lesbian, and bisexual Australians and correlated generally with sexual adventurousness. Researchers concluded, "Our findings support the idea that BDSM is simply a sexual interest or subculture attractive to a minority, and for most participants not a pathological symptom of past abuse or difficulty with 'normal' sex."[74] The lead researcher additionally remarked that other Australians likely also participate in mild BDSM activities, but are reluctant to identify with a label they perceive as having negative connotations.[75]

BDSM, then, is a somewhat risky but potentially rewarding activity to which adults can consent in the same way they might consent to alcohol use or rough contact sports. Anti-rape activist Stacey May Fowles takes the issue of consent a step further, however, by arguing that BDSM sex is actually *more* consensual than conventional "vanilla" sex. She writes:

> [T]he very existence of a safe word is the ultimate in preventing violation [. . .] at any moment, regardless of expectations or interpretations on the part of either party, *the act can and will end.* Ignoring the safe word is a clear act of violation that is not up for any debate. [. . .] The safe, sane, and consensual BDSM landscape is made up of stringent rules and safe practices designed to protect the feelings of *everyone* involved, and to ensure *constant, enthusiastic consent.*[76]

Mainstream sex rarely has any structured method by which consent is obtained, affirmed, and withdrawn: to put it bluntly, vanilla sex has no safeword. As Fowles points out, however, practices such as safewords require a community that will sustain and enforce these norms, and the community must not grow so fast that newcomers cannot be socialized into it. A friend tells me of a kink party he attended with a girlfriend. An acquaintance approached the girlfriend and started to tickle her, and she playfully cried, "Safeword! Safeword!" Not only did the tickling immediately stop, but all activity at the party ceased as if someone had thrown a switch (much to her embarrassment, as she hadn't been using the safeword in earnest). This incident demonstrates the seriousness with which that community takes safewords—if her use of the safeword had been real, she and her play partner would have immediately been offered support and aid by multiple attendees. But it also demonstrates the necessity of socializing newcomers into community norms. The young woman's embarrassment taught her a lesson about respecting the practices that create a safe space for BDSM play. Yet that lesson would not have been as effectively learned if the party had been primarily attended by newcomers with a similarly shaky understanding of the rules. (For more discussion of alternative erotic communities, see Chapter 5.)

Some feminist thinkers continue to argue that the apparent consent of female BDSM practitioners is a result of patriarchal brainwashing, and it has even been suggested that patriarchal oppression generally compromises women's consent to heterosexual sex almost to the point of collapse. As Catharine MacKinnon states in a 2006 interview:

> The [widespread cultural] assumption is that women can be unequal to men economically, socially, culturally, politically, and in religion, but the moment they have sexual interactions, they are free and equal. That's the assumption—and I think it ought to be thought about, and in particular what consent then means. It means acquiescence. It means passivity.[77]

For MacKinnon, enthusiastic consent to heterosexual sex is impossible because the conditions surrounding sex are inherently coercive. The idea that in patriarchal culture, all heterosexual sex borders on, or actually is, rape[78] has been firmly rejected by most progressives, but I feel the idea is worth returning to because it has a grain of truth. It is true that we live in a web of unequal, often exploitative power relationships in which women still frequently hold the shorter end of the stick. A woman who does not have the same economic, political, and social power as her partner is at a disadvantage if she wishes to leave the relationship. Even within my highly progressive community, it is common for heterosexual couples to put off divorce because the woman is dependent on her spouse's health insurance or does not earn enough to afford her own living space. Such conditions put pressure on the disadvantaged partner to engage in sex that she may not want.

To say that an adult person with access to education, jobs, and basic political freedoms *cannot* give consent to sex, however, proactively places such individuals in the role of victim. Rather than empowering the disadvantaged by encouraging them to maximize and increase their personal and social agency, it threatens to strip them of whatever power they do have and invalidates their expressions of desire. Further, to declare that sex in a pressured environment is automatically nonconsensual blurs the helpful line that Nielsen draws between exploitation (where vulnerability is taken advantage of) and coercion (where the use of force removes the possibility of consent). In a situation with an unequal power dynamic, the party with more power has moral responsibility for exploiting another's vulnerability. However, so long as the exploited party retains the ability to give valid consent, he also has the ability and responsibility to withdraw it and/or to leave the relationship. In an exploitative relationship, moral responsibility is shared between both parties, not assigned wholly to the perpetrator as it would be in the case of nonconsent or coerced consent.

There are some cases where coercion operates systemically. The circumcision of young girls (aged four to eight) by Sudanese women serves as one sad example of consent invalidated by a social system. The procedure, called Pharaonic circumcision or infibulation, removes the clitoris and a large part of the labia, then fuses the wound so that the resulting vaginal opening is no bigger than a pinhole. The resulting mass of scar tissue acts as a kind of chastity belt, preserving the "sexual purity" of the girls and the honor of their families. The vaginal opening must be painfully re-opened after marriage, sometimes with a sharp instrument.[79] The procedure is technically illegal under Sudanese law, but the law remains inconsistently enforced. Although Sudanese women give consent on behalf of their daughters to undergo this procedure, which is often performed by women (midwives or other medical professionals), their consent is partially, if not wholly, invalidated by their limited access to education and independent financial resources. Many Sudanese women may only be educated, venture out in public, or hold jobs under the strict supervision of their husbands. In 1989, researcher Hanny Lightfoot-Klein found that Sudanese women were generally shocked to learn that infibulation was not a universal human practice. Since a girl who is not infibulated is unmarriageable, Sudanese women cannot freely refuse the procedure on behalf of their daughters without imperiling their daughters' social and economic future. Further, as many are unaware that the procedure is not universal nor religiously required of Muslim women, their consent is not informed. Under these circumstances, consent to infibulations is coerced and therefore invalid.

In the absence of such systemic coercion, however, it is imperative to respect the right of adults to control their bodies and to engage in consensual acts. To do otherwise disrespects individual and cultural differences and needlessly restricts individual autonomy. In parts of the world where women enjoy political freedoms, an educated, professional Muslim or Jewish woman may choose to cover her hair as a principled religious choice. Similarly, an adult American woman who seeks out BDSM play is not necessarily a victim of cultural pressures. Her desire for BDSM activities puts her at risk for social disapproval and, depending on her social circumstances, may endanger her job or her custody of her children. Such a person must be willing to challenge multiple social norms in order to fulfill her erotic desires. In the absence of obvious coercion (and assuming that she expresses satisfaction with her choices), such nonconformist behavior is best read as an expression of personal autonomy.

Although the material provided here is too brief to fully explore all the nuances of consent issues, I offer it as basic context for the discussions of touch and sexuality that follow. In a positive theology of the erotic, no consensual erotic act between adults is fundamentally wrong, although some acts may be unwise. It may also be unwise to quit one's job and travel the world looking for the meaning of life or to take up drag racing as a hobby, but these activities are hardly unethical in themselves. In an erotic ethics that values personal autonomy, individuals have the right to experiment with their lives and make their own mistakes. Admittedly, an erotic ethics that is aware of interdependence also calls us to consider the impact on loved ones if we come to harm. Yet concern for loved ones and community must be balanced with the knowledge that it is emotionally toxic to live a closeted and inauthentic-feeling life. Attending to the impact of our actions on others does not mean treating them as though they cannot handle our true selves, or refusing them deep intimacy because we fear they cannot bear to see the sources of our joy. Within this ethical framework of interdependent community and personal autonomy, the ability of adults to consent to erotic contact should be absolute.

Overview

Chapter 1 addresses the nature of the erotic on the broadest possible level: the cosmological. Engaging The Charge of the Goddess by Doreen Valiente, the Anderson Faery tradition myth of the Star Goddess as told by Starhawk, and writings by a variety of other poets, artists, and theologians, I explore the erotic as a fundamental aspect of human experience that opens us to our divine nature and to connection with others and the world. Though initially articulated through myth and poetry, the erotic emerges as a theological principle that can ground a system of ethics.

Chapters 2 and 3 delve into social ethics. Since, as Marvin Ellison writes, every act of social exploitation involves a violation of the body, valuing pleasure as a human right potentially increases social cohesion and justice. Chapter 2, therefore, looks at the functioning of the erotic in the social and political realm. Ellison argues for eroticizing equality, rather than eroticizing hierarchical power differentials the way mainstream society currently does. I instead advocate eroticizing *empathy* to avoid confusing social equality with social equivalence. The writings of sexual minority activists, bodyworkers, and both Pagan and Christian theologians are engaged to explore the erotic on a political level.

Chapter 3 argues for the necessity of a pluralistic erotic ethics that understands erotic diversity as beautiful, normal, and holy. Playing off Gayle Rubin's seminal essay "Thinking Sex" and liberation theology's idea of a "preferential option for the poor," this chapter asks what unique gifts of perspective and wisdom sexual minorities might offer to the mainstream. Pornography, BDSM, polyamory, transgender, and adolescent sexuality are treated as sites of political and ethical controversy that offer insights into the broader nature of human sexuality. Special attention is given to models of sacred sexuality in the Pagan movement (successful and unsuccessful).

Chapter 4 narrows its focus from the social realm to consider the erotic on a personal level. Drawing on the spiritual and clinical writings of bodyworkers and on contemporary Pagan practices of sacred and/or healing touch, I present sacramental touch as a practical expression of belief in an erotic cosmos: a religiously grounded practice of mindful, intentional touch between individuals. Deane Juhan, David Lauterstein, Ashley Montagu, and Phyllis Davis serve as conversation partners in providing a philosophy and guidelines for how readers might practice sacramental touch in their ordinary lives. Studies of the effects of touch deprivation on animals and children are used to show the developmental impact of a failure to value loving touch and physical pleasure.

Chapter 5 confronts the dangers of treating touch as a sacrament. In addition to acknowledging sexually transmitted diseases and the risk of sexual assault, I examine the medical industry's role in covering purity taboos around touch with apparently medical models of hygiene and disease. Additionally, the chapter considers potential psychological and social risks of practicing sacramental touch. Again, the efforts of Pagans and sexual minorities to support alternative systems of erotic ethics through community accountability are given special attention.

In the Conclusion, the importance of erotic theology is framed by Sherry Turkle's recent book on technology and human intimacy. In conversation with the animist theology of David Abram, I suggest that the conscious experience of being embodied—of touching and being touched—is necessary to make us human, for biological as well as spiritual reasons. For Pagans and other religious people, sacramental touch can be a concrete expression of a holistic, body-centered ecotheology.

CHAPTER 1

Divided for Love's Sake

AN EROTIC COSMOLOGY

When I was in college, my partner and I once stopped for a chat with our Shakespeare professor, a dynamic and funny lecturer who was able to identify—convincingly!—homoerotic overtones in every play we read. The conversation turned to recent films, and I mentioned having seen Cronenberg's 1996 film *Crash*, a surreal tale of young people who deliberately crash their cars in order to become sexually aroused.

> "Was it erotic?" my professor asked.
> "Well, it had a lot of sex," I answered.
> "Yes, but was it *erotic*?" he asked again.

I suddenly grasped his meaning and realized that I had never really understood the word. In class, as we studied Shakespeare's endless parade of cross-dressing characters and their loves, I'd always heard the word *eroticism* as meaning that these charged relationships were, somewhere deep down, all about genitals. Yet my professor's question alerted me to something I hadn't seen: the characters in *Crash* might be employing their genitals left and right, but that didn't necessarily mean there was anything erotic about it. *Eros*, then, had to be something else—intimacy, a feeling that engaged the emotions as well as the body, something that could move between two Shakespeare characters whose only physical contact might be a clap on the shoulder. In a twisted way, the encounters of the car-crashing young people in Cronenberg's film are fumbling toward the erotic: their engineered brushes with mortality spark an intense desire for human connection, a desire that highlights the alienation of their daily lives and their ignorance of how to relate to each other.

In American culture, touch of any kind is frequently sexualized, to the extent that many adults only receive affectionate touch from their romantic partners. Indeed, some of us seek out sex not because of specifically sexual desire, but just out of desperation to make contact, emotional and physical. Ben Benjamin and Cherie Sohnen-Moe put it bluntly in *The Ethics of Touch*: "A person who only feels intimate with someone when sex is involved might start to believe that intimacy and sex are the same thing."[1] This confusion of the sexual with the larger categories of intimacy and eroticism contributes to a touch-starved society. As a massage therapist, much of my work's therapeutic impact comes from the fact that my clients simply do not receive enough pleasurable touch. This physical and emotional deprivation leads to increased levels of stress and anxiety. Providing ethical therapeutic touch, however, requires constant policing of the boundaries of the therapeutic container. To prevent my touch from being misunderstood as sexual, I employ a host of consent and release forms, and I drape the client's unclothed body with sheets. I dress modestly and observe myself carefully while working, lest I accidentally brush a leg against my client's dangling hand. And yet, as I was advised in school, these basic tactics may not be enough. One instructor suggested that, regardless of my marital status, I might wear a wedding ring when meeting male clients for the first time; the assumption that I'm married might head off thoughts of sexual contact on the client's part.

As a massage therapist, I find that sex and massage have little in common—I spend a great deal of time visualizing my clients' individual muscles, as if they had no skin! Yet the risk that my connected, compassionate touch might be taken as sexual is significant. My touch *is* erotic, a natural outflow of my desire to comfort and heal. My work suffers and becomes less effective when I am not open to connection with a client, when I do not allow myself to sense the patterns of tension and pain that are carried in a client's body. Whether I am giving a massage or merely shaking a client's hand, my willingness to open to the erotic makes all the difference in the quality of my contact, the warmth of my handshake, the frankness with which I look someone in the eye. My openness to the erotic determines everything about how present I can be in the moment and in my own body.

Defining the Erotic

What, then, *is* the erotic? I've spoken of it so far as a quality of connected and intimate touch, one that can include the sexual but is not limited to it.

Further, it is a quality that disconnected sexual expression utterly lacks. For me to violate a client's boundaries with an inappropriate sexual comment would be completely *un*erotic—disturbing and potentially traumatic to the client, who trusted me enough to relax and become vulnerable on my table.

Early in the twentieth century, psychologist C.G. Jung defined eros as a relational principle, "the great binder and loosener" within the psyche, among people, and between the individual and the world.[2] Because of his historical time and place, Jung considered personality traits to be either masculine or feminine in nature, though he believed that both kinds of traits were present in every individual. In Jungian psychology, the masculine and the feminine sides of the personality must be developed and integrated for a person to achieve maximum emotional and mental health. Today, as gender roles become more fluid in our society, many people are inclined to see the desire for relationship simply as a *human* trait, not as a feminine one as Jung might have. Jung broke important ground, however, when he saw the erotic as the desire for wholeness within the self, as well as the desire to connect and interact with others. Nor is eros simply a human principle. In Jungian thought, the human self reflects the *imago dei*, the image of God in which humanity is created. The desire for wholeness within and among human beings, then, is an extension of divine erotic desire. Eros is a cosmic principle, not just a human one—a force active in consciousness itself.

Speaking of the erotic as a psychological force, however, may distract us from the fact that it must be embodied. Much as human beings love to speak of transcending the body, even the thought of transcendence arises from the spongy grey matter of our ever-so-physical brains. Our bodies are the vehicles with which we experience all of Being. Accordingly, the erotic is an urge toward wholeness and connection that both engages with and delights in flesh. In her essay "Uses of the Erotic: The Erotic as Power," poet and feminist Audre Lorde describes eroticism as the impulse to flourish, a desire for pleasure that drives a person to resist oppression and strive toward a joyful, satisfying, and meaningful life. She writes:

> [One] way in which the erotic connection functions is the open and fearless underlining of my capacity for joy, in the way my body stretches to music and opens into response, harkening to its deepest rhythms so every level upon which I sense also opens to the erotically satisfying experience whether it is dancing, building a bookcase, writing a poem, or examining an idea. [. . .]
>
> This is one reason why the erotic is so feared, and so often relegated to the bedroom alone, when it is recognized at all. [. . .] In touch with the erotic, I become less willing to accept powerlessness, or those other supplied states of being which are not

> native to me, such as resignation, despair, self-effacement, depression, self-denial.[3]

For Lorde, the erotic can be expressed sexually in the bedroom, but it is also present in every moment of intense engagement with the world, particularly moments that are celebratory or creative. Where eroticism flows, the tendency to become alienated from one's environment is diminished; experiences of joy and pleasure contribute to valuing oneself in a way that reduces tolerance for mistreatment and abuse. To experience the erotic, suggests Lorde, is to know the capacity for joy as a human birthright, as a gift of our physicality. By simply being born, we inherit the capacity for connection.

Lorde sees the erotic as a quality of living that infuses every meaningful life. Her description is limited to the human realm, however, while like Jung, Christian theologian Carter Heyward sees the erotic as a cosmic principle of connection. "My eroticism," she writes, "is my participation in the universe."[4] Any relationship, whether with another individual, a favorite natural setting, or a divine presence experienced in ritual or prayer, is potentially erotic when based on connection and a longing for intimacy. Yet the physicality of the erotic is no less potent when experienced without direct human contact. Following the conventions of her Craft tradition, esoteric writer T. Thorn Coyle employs the term "Sex" for the erotic, but her description emphasizes its presence within and among humans and their environment, not just between people:

> Sex is the rush of sap into the tree [. . .] It is the caress of bee leg on rose petal, the rooting of the mole into the earth. It is the line of my flesh etched into space by my breath. It is the line of your flesh etched by my hand. It is the heartbeat of the earth and the sun raising sweat beads on my skin [. . .] Sex is connection to all of life.[5]

Similar to Coyle's definition is that of writer and literary critic Miriam DeCosta-Willis, who calls eroticism "The life force that flows like an inscrutable tide through all things, linking man to woman, man to man, woman to woman, bird to flower, and flesh to spirit. [. . .] Desire. Pleasure. Wholeness."[6]

Coyle, deCosta-Willis, and Heyward experience Being as soaked with the erotic. Yet to feel the pleasurable interpenetration of all that is requires a shift in awareness. Such a shift may be as subtle as tuning a radio to the right station, but it can sometimes require a temporary separation from old patterns of disconnected living. In *The Spiral Dance*, Pagan theologian

Starhawk writes of a summer she spent as a young person, bicycling up and down the West Coast and camping on beaches:

> For the first time, I lived in direct contact with nature, day and night. I began to feel connected to the world in a new way, to see everything as alive, erotic, engaged in a constant dance of mutual pleasuring, and myself as a special part of it all.[7]

To experience the erotic as a cosmic force gives everyday life the emotional flavor of a parent cradling a baby, an embrace between long-separated friends, or lovemaking with a trusted partner. To feel the intake of air into one's lungs as a caress from the universe opens one to a presence whose capacity for nurturing is as vast as the earth itself.

An Erotic Creation Myth

Pagan, Jewish, and Christian theologians of the erotic frequently understand it as the principle from which all of Being springs. Quoting Jewish theologian Martin Buber, Heyward asserts, "In the beginning is the relation."[8] We are born into relationship, without which—as helpless infants—we cannot survive. A child comes to know herself in the arms of her caretaker, in that first moment when she perceives herself as separate from the person tending to her needs. The knowledge that one exists separately from others comes simultaneously with the knowledge that there are others to whom one is relating. Yet beyond this human reality is the cosmic notion that erotic relationship brings the universe itself into being. This is the mystery expressed by the creation myth of the Anderson Faery tradition of the Craft, a myth beautifully retold by Starhawk in *The Spiral Dance.*

> Alone, awesome, complete within Herself, the Goddess, She whose name cannot be spoken, floated in the abyss of the outer darkness, before the beginning of all things. As She looked into the curved mirror of black space, She saw by her own light her radiant reflection, and fell in love with it. She drew it forth by the power that was in Her and made love to Herself, and called Her "Miria, the Wonderful."
>
> Their ecstasy burst forth in the single song of all that is, was, or ever shall be, and with the song came motion, waves that poured outward and became all the spheres and circles of the worlds. The Goddess became filled with love, swollen with love, and She gave birth to a rain of bright spirits that filled the worlds and became all beings. [. . .]

> All began in love; all seeks to return to love. Love is the law, the teacher of wisdom, and the great revealer of mysteries.[9]

In this primordial moment, creation happens through an erotic encounter. But is Miria a part of the Goddess, or a separate being? Is this the Big Bang visualized as an act of self-pleasure, or one of lesbian lovemaking?

As is often the case with mysteries (used in the religious sense of "truths that must be experienced to be understood"), the best answer to these questions is "yes, and no." In this cosmology, the Goddess is not outside creation, but *is* creation. Miria begins as the Goddess' reflection, seen in a mirror of nothingness. She is then "drawn forth" from the Goddess, but we are not told the nature of the shift—whether it occurs physically or merely on the level of perception. (When approaching cosmology on this poetic level, the distinction between the two blurs.) The Goddess divides herself for the sake of love and pleasure, but Miria, being the Goddess' mirror image, maintains the same nature as the Goddess. Yet now the nature of the Goddess can begin to be known. Alone in the darkness, the Goddess had nothing with which to compare, no mirroring Other by which to know her Self. As with an infant, before the moment that the caretaker is recognized as separate, there is no self-consciousness, no difference between "I" and "You" (or, to follow Martin Buber's tender and reverent phrasing, no difference between "I" and the cherished "Thou," the perpetually beloved Other). Until the Goddess splits herself into Miria, She is nothing, simply the undifferentiated blackness of endless potential. Conscious being, suggests the myth, is inherently plural.[10]

In *The Book of the Law*, a channeled document recorded by influential twentieth-century magician Aleister Crowley, the primordial star goddess Nuit declares, "I am divided for love's sake, for the chance of union."[11] The Goddess creates Miria from Herself, but Miria's creation appears to be an act of division, not one in which new substance is created. The Goddess has taken a part of her body and given it a new name; the boundary between herself and Miria is a constructive illusion, one that allows pleasurable interaction. As in *The Book of the Law*, the division occurs *for love's sake*. A space is opened between the lovers so that desire, longing, and the pleasure of reunion become possible. From this primordial erotic connection and the ecstasy it generates, all creation is born in an orgasmic "rain of bright spirits"—all pieces of the divine fragmented for the opportunity of interaction, self-knowledge, and pleasure. We might see animals, plants, the gods, and ourselves as these bright spirits filling the worlds—pieces of divinity, yet simultaneously whole unto ourselves, just as Miria is the Goddess' complete reflection. Though we have been divided from the whole, this too is an illusion made for love's sake. In our experiences of each other and the world, the divine in us and through us sees itself more clearly, each of us as cherished as the perfect lover or a child newly born.

The erotic, then, is a force that operates within the self and between selves, just as the Goddess is both the whole of creation and all of its constituent parts. The tender and passionate desire we feel for our loved ones is also something that can move within us, softening our judgment of our perceived flaws and allowing us to treat ourselves with the same compassion we would a child. This love for the self is part of the resilient joy of which Lorde speaks—to allow the erotic to move within us, particularly in potentially solitary activities such as creative work, dance, prayer, or self-pleasure, is to be open to knowing ourselves as both lover and beloved. Needing no external force to experience our own worthiness of love, we find an internal ground on which to stand.

Starhawk's creation myth is told as a myth of a Goddess, and particularly when *The Spiral Dance* was first published in 1979, potent images of the divine as female served as important sources of inspiration for women in the feminist movement. In her hands, the myth is told against the creation myth in the book of Genesis, with its apparently masculine God: "The world is born, not made, and not commanded into being," she writes.[12] Starhawk notes that the Goddess "could just as easily be called *God*,"[13] as gender does not yet exist. To understand the creating deity as female, however, emphasizes the nature of the creation process as one of gestation and birth: organic, internal, nurturing, and the result of erotic energy. Yet the term *Goddess* also tends to obscure the pre-gender nature of the deity, for whom maleness, femaleness, and an infinite range of non- or alternately-gendered ways of being are possible. Poet and Craft teacher Victor Anderson, whom Starhawk credits as her source for the myth, calls this deity "the clitorophallic God Herself"—a term meant to encompass the divine masculine, the divine feminine, and the divine hermaphrodite.[14] God Herself is not without gender, but instead contains the potential for all genders. Self-stimulating, both yonic and phallic, God Herself births Being using only what she contains within hir. Though English lacks good language to describe genders other than female and male, the word "clitorophallic" is provocative. It suggests a state in which gender *must* be more than anatomy, because the anatomy of the Goddess is the body of the universe itself: too complex and multiform for gender as we usually understand it.

Yvonne Aburrow similarly demonstrates the importance of gender-bending and androgyny in other contemporary Pagan traditions, particularly in the gay men's movement called the Radical Faeries, but also in traditions that use conventional images of gender, such as British Wicca and Heathenry. According to the nineteenth-century occultists whose work ultimately influenced the Pagan revival, psychological androgyny was a key spiritual goal. As Aburrow argues, this theme remains in the theology of many contemporary Pagan traditions, including Wicca (though it is not always fully acknowledged). Despite British Wicca's apparent focus on conventionally gendered heterosexuality, one central Wiccan prayer is addressed to an androgynous or hermaphroditic entity characterized as "male and female, the source of all things"[15]—an image similar to Starhawk's pre-gender "Goddess."

Such images have a long history in Western mystery traditions. For example, a multi-gendered image of primordial deity appears in the Gnostic traditions of Christianity. According to early Christianity scholar Elaine Pagels, the two creation stories in Genesis were both read by Gnostics as involving an original, androgynous human being. The first creation story in Genesis, in which God speaks creation into being, was understood as referring to an androgynous deity. Humanity, created in the image of the God, is "male and female," and so must God be as well. In Gnostic readings of the second creation story, in which Eve emerges from Adam's side, Adam is "'discovered to be two,' an androgyne who 'bears the female within him.'" God and God's creation are understood to be "bisexual" or "masculo-feminine."[16]

The strangeness of terms like "masculo-feminine" points to a truth that our language is ill-equipped to express—there are other orientations to the erotic beyond the polarity of maleness and femaleness expressed in sexually-reproducing species. Starhawk suggests:

> Sexual reproduction is an elegant method of ensuring maximum biological diversity. [. . .] But to take one particular form of sexual union as the model for the whole is to limit ourselves unfairly. If we could, instead, take the whole as the model for the part, then whomever or whatever we choose to love, even if it is ourselves in our solitude, all our acts of love and pleasure could reflect the union of leaf and sun, the wheeling dance of galaxies, or the slow swelling of bud to fruit.[17]

As discussed in the Introduction, the term "queer" has come to express an orientation to sexuality that does not accept heterosexuality as the norm. Loosely applied, the word embraces sexual minorities and their allies as a coherent community, one that is committed to diversity in erotic expression. Expanding the notion of "queer" beyond the human, beyond sexuality into the cosmic, makes sense of Starhawk's claim that the "wheeling dance of galaxies" is as intimate and sensual as lovemaking. We live in a queer universe, in which erotic connection is possible between any two—or more—beings or energies. Gender or the details of physical anatomy are not barriers to eroticism. The universe, God Herself, reaches out to all hir many forms in love, bridging the gap between plant and animal, water and sun.

The Transgressiveness of Divine Eroticism

For some working in Pagan and Western esoteric traditions,[18] the intensity of cosmic eroticism can only be captured with graphic and potentially transgressive sexual imagery. One example is British writer and graphic novelist Alan Moore, a practitioner of Western esotericism who also cites the Thelemic philosophy of esotericist Aleister Crowley as a primary influence. Crowley's work is often perceived as problematically sexist and warped by his traumatic, puritanical upbringing in a conservative Christian sect. Nevertheless, his embrace of sexuality as a source of personal and spiritual empowerment—and his particular celebration of women's sexuality—continues to resonate with Pagans and other alternative religious practitioners.[19] As women's social and political freedoms have expanded, feminist women and men have approached Thelemic philosophy and Crowley's visions of erotic feminine power as tools for personal liberation, and in some cases, for empowerment and healing after sexual assault.[20] Similarly, Alan Moore—who is an outspoken advocate of LGBT rights,[21] and whose works often champion strong female characters—seeks to read Crowley against his patriarchal context. Like many feminist Thelemites today, Moore uses Thelemic imagery to sanctify sexualities traditionally regarded as transgressive.

Although it is superficially constructed as a superhero narrative, Moore's graphic novel series *Promethea* is read in Pagan and occult communities as an instructional text.[22] In addition to lessons on Tarot (a system of divination) and the neoplatonic elements (earth, air, fire, and water), the series includes a detailed exploration of the hermetic kabbalah, a metaphysical system originally derived from Jewish mysticism. In the midst of this kabbalistic exploration, the main characters encounter a traditionally problematic image of feminine power in a new light. Rendered with heavy black lines reminiscent of medieval or Renaissance woodcuts, the Whore of Babylon (or Babalon, in Moore and Crowley's preferred spelling) appears jeweled and voluptuous, riding a beast as she does in the biblical book of Revelation. For many Christians, the Whore is the personification of carnal sin. Moore, however, sees her as pointing to a truth about the nature of humanity's relationship with the universe.[23] "When men KNOW her, she is then called REVELATION, for she lewdly shews herself," explains the Elizabethan magician in Moore's text.[24] The mystical unveiling of the divine nature of Being is compared to the shameless nudity of a prostitute, inviting all to come and experience her intimately. Then Babalon speaks for herself:

> About the markets of the world I am sent bare, until the worst of creatures may lie down with me. Though wretched and with sores, they shall not be refused. [. . .] I am She no earthly Man or Woman may embrace, that yet is Whore to ALL.

No one, no matter how poor of body or spirit, no matter how dirty, diseased, or broken, is refused the erotic love of the Goddess, who is the face of Being itself. Moore suggests that this is the meaning of the sphere of divine energy in which the encounter takes place: it is called "Understanding," a translation from the Hebrew word "Binah." Understanding—a deep understanding that does not shy away from hard truths, but reveals each person exactly as she is—is portrayed as a whore, because such understanding holds nothing back.*

Daily life is indeed often messy and painful. We have all, at times, felt unworthy of being loved; some of us have experienced the kind of systematic humiliation and degradation that Lorde speaks of. Yet to contemplate the Goddess as a whore is to imagine a divine embrace that does not shy from our imperfections—not even from those diseases (such as HIV or herpes) that make people into social pariahs, perceived as both physically and morally "impure." Through Babalon, Moore suggests that the universe desires us just as we desire her, profoundly and erotically, and that she experiences great pleasure in her union with and knowledge of us. This Whore is not a disadvantaged teenage sex worker, forced into prostitution in order to support herself. Rather, she embodies the energy of the temple prostitute, a myth that may not be based on any historical religious practice, but nevertheless expresses a human desire to be unconditionally loved: not just as a mother would, but as only someone who sees all our corruption can. Moore presents the erotic longing of God Herself for humanity with the symbolism of transgressive sexuality. This love is not about genitals—or at least, not *only* about that one sacred part of ourselves—but about a love that holds back nothing and is intense, carnal, and shameless in its desire.

As interdisciplinary sexuality scholar Loraine Hutchins points out, when applied to human beings, imagery of the Whore can be problematic. "Sacred

* Portions of this discussion are revised from Christine Hoff Kraemer, "The Undying Fire: Divine Love and the Erotic in *Promethea.*" In *Sexual Ideology in the Works of Alan Moore: Critical Essays on the Graphic Novels* © 2012 Edited by Todd A. Comer and Joseph Michael Sommers by permission of McFarland & Company, Box 611, Jefferson NC 28640. www.mcfarlandpub.com

whore" imagery has recently been embraced by sex-positive feminists advocating for spiritual feminist and/or Pagan approaches to sex work. Integrating counseling techniques such as conscious touch, communication training, and relationship coaching with tantric and other religious approaches to sexuality, these women (and a few men) present themselves professionally as "erotic healers" available to help clients deal with sexual wounds. Through dialogue with feminist sex workers, organizations such as the National Organization for Women have become more open to the possibility of feminist approaches to sex work and of unionizing sex workers. Hutchins supports the therapeutic work of these modern-day "sacred prostitutes" and recognizes that some women associate sacred whoredom with spiritual authority, as well as with control over their own sexuality.[25] However, she emphasizes that sacred whore imagery must be presented in the context of a larger political and social movement for sexual and gender equality, including a focused critique of the sex industry.[26] Accordingly, it is important to understand that Moore's use of Babalon as a Goddess figure occurs in the wider context of his gender-egalitarian, sex-positive philosophy. For Moore, Babalon is a positive, if somewhat intimidating, figure.

Moore's representations of deity are attempts to poetically represent the nature of the nonhuman universe with an accessible anthropomorphic image. Western esoteric philosophy asserts that "As above, so below; as within, so without," suggesting that patterns repeat themselves on multiple levels of Being. The microcosms of individual human selves are considered to be mirrors of the macrocosm of the universe, and vice versa. This truth is poetic, not literal, however; if a human being is a microcosm of the universe, it does not follow that the universe has a physical or psychic anatomy exactly like a human's, only larger. Additionally, it is tempting to put the human being in a special position as "the" microcosm and to conclude that the universe functions along human principles. Yet there are many microcosms that the universe as a whole reflects—not just the human, but the atom; not just the earth, but vast nebulas. This complexity is expressed in our particularity, but we remain connected to, and can potentially access, an ineffable awareness of the whole. Speaking through the mouth of one of his characters, Moore writes that life—and especially humanity—is "the clay in which the forces that shape all things leave their fingerprints most clearly."[27] The conscious human mind may be too limited to contain the universe, but we still open to the whole in moments of ego-blurring ecstasy: in religious ritual or during sex, or perhaps no special circumstance at all, but just a sudden connected moment in the midst of ordinary life. Though it is difficult to speak of these experiences, poets and religious writers have done their best to craft words pointing us toward them—their writings and teachings being not the experience itself, but rather a map encouraging us to seek out the territory.

To know that we are a microcosm of the macrocosm allows us to trace the movement of the erotic on many levels and sense a sacred wholeness in daily life. The erotic is the gravitational attraction of planet to star, the caress of salt water on a sandy beach; it is the pulsing energy of a crowd dancing at a rock concert and the electricity of a first kiss; it is a dog rolling ecstatically in fresh grass; it is a salesperson's smile when it genuinely touches her eyes. Yet the erotic also moves within us. It is my compassion toward myself when my body is tired and sore, and I stop work for a hot shower and a nap. It is my willingness to sit with anger or jealousy, listening to what they say about my boundaries and needs rather than pushing them to the back of my mind. It is the ability, in the midst of struggle, to give myself the same nurturing advice I would give a dear friend or lover—and to act on it. The gods move in me, reflecting each other, making love, birthing new realities. I am both a piece of God Herself and a mirror of the whole. The erotic churning of the universe—fragmenting, interacting, reuniting, dividing again—moves us towards greater self-knowledge: not a purely intellectual knowledge, but a bodily and experiential *gnosis* that is deeper and more integrated than can be expressed through words. Our fumbling for awareness is part of the larger process of divinity doing the same.

Pleasure as Biological Strategy

Lest we lose ourselves in trying to envision the cosmic, however, let us remember that the erotic is grounded in our immediate physicality. Bodyworker Deane Juhan emphasizes the concreteness of the erotic in human life when he characterizes the existence of pleasure as an evolutionary strategy. He cites studies on the ways that stressful conditions lay the groundwork for disease and psychological dysfunction, while relaxation and pleasure enhance the body's immune response and contribute to improved mental and emotional health. He writes:

> Now this is exactly where sensory "pleasuring" derives its survival value: It is one of nature's key antidotes to these pernicious developments [the consequences of stressful conditions on the body], and this is why evolution has so highly developed our sensual capacities for pleasure in the first place. The mind *attends* to pleasure, is inclined towards it, opens its listening channels in order to amplify it, strives to accommodate and sustain it. It is *hedonic.* And through these expanded portals of positive affect come

> streaming clearer, more complete, and more detailed sensory *information*—without which the finest brain cannot successfully direct internal biological traffic or external behavior.[28]

Juhan suggests that due to evolutionary pressures, when organisms behave in ways that are associated with health and thriving, pleasure is the result. Additionally, the relaxation and sensory awareness that come with pleasure provide an opportunity for the organism to receive and process complex information about itself and the environment, information that is necessary for the organism to regulate itself on both conscious and involuntary levels. One simple example of this phenomenon is one that is probably familiar to many readers: at the end of a long, busy day, I often come home, take off my shoes, and settle on the couch with my feet up to rest and breathe. More often than not, in this moment of relaxation, there is a sudden rush of bodily knowledge: I'm hungry (maybe I even forgot to eat). My back was starting to tense up, but my heating pad is doing its job and feels delicious—probably preventing a spasm a few days down the line. Moments of pleasure interrupt the fight-or-flight state many of us find ourselves in on a daily basis and allow our bodies to finally prioritize the need for rest and maintenance. Not only does stress often cause us to become unaware of appetite and other bodily signals, but while in a highly stressed state, biological processes such as digestion and immune function are partially suppressed. Pleasure and relaxation allow these necessary processes to resume. Juhan continues:

> Any organism that can surrender its defensive armoring, attend to genuine biological (as opposed to merely socially acculturated) pleasure, and integrate the information that comes along with it has the capacity to self-regulate towards a continually developing optimal. [. . .] Eros in its widest and most significant dimensions means the hedonic pleasure associated with surrender to *all* of the pleasure-giving physical and mental processes that genuinely contribute to the vitality of our internal life and contribute to our survival as individuals and as a species.[29]

As Juhan suggests, since our environment is a complex system, it is possible for the particularities of a pleasure-seeking strategy to become outdated or confused. Most human beings love the taste of fats and sweets—an excellent characteristic in a hunter-gatherer society when such calorie-dense foods could be hard to come by and were also the best possible support for a physically demanding existence. In contemporary society where such foods are ubiquitous and the work that many people do involves a great deal of sitting or standing, however, it is possible to eat a diet consisting mostly of fats and sweets, with considerable negative impact on one's health. Particularly when conditions are stressful and an individual is deprived of other forms of pleasure—such as touch, physical activity, or rest—the craving for pleasure can become unhealthily centered on a single, easily available source, such as food. Accordingly, Juhan warns against socially enculturated patterns of pleasure-seeking that ignore the holistic needs of the body. A weak, overweight body that receives little touch is a body in pain, and the chronic ailments of our culture—tension headaches, low back pain, frequent minor illnesses, etc.—attest to the fact that pleasure can be sought superficially, in a way that does not address pain's underlying causes. A combination of limited choices (or a perception of limited choices) and social conditioning can encourage this kind of self-defeating behavior.

When Juhan speaks of "hedonism," however, he intends a survival strategy in which organisms, given a wide range of choices, are led by the experience of pleasure toward a state of optimally balanced health. (Juhan's use of the word "surrender" above suggests that this process cannot be a wholly intellectual one, but requires an intuitive, sensual, and holistic approach to the body's needs.) A relatively simple mechanism produces extremely complex behaviors. Organisms gravitate toward experiences of pleasure, sensual as well as cognitive. We enjoy food and sex in a way that is qualitatively different from the way we might enjoy drawing a picture or watching a thoughtful film, although both kinds of experience are recognizably pleasurable. When an experience pleases us, we tend to seek out more of it. In a context of many choices, however, the desire for a particular pleasure is generally self-limiting. Most of us do not stay up for days without eating to draw pictures until our hands are cramped and sore; even when swept up in a creative frenzy, we do not work ourselves to a state of literal starvation, or lose all interest in calling our friends, having a hot bath, and going out dancing. When we perceive that pleasure is restricted to one or two sources—such as food or a drug—we sink into addiction.[30] This is not a total perversion of the pleasure-seeking survival strategy, but an organism's best attempt to thrive in scarcity. If the only reliable pleasure in life comes from a drug, to repeatedly turn to it still demonstrates a commitment to life, rather than the suicidal detachment that comes from thinking that no pleasure is possible. In some ways, this kind of addiction is easier to untangle than the compulsive behavior that comes from experiencing only superficial and fleeting pleasures. Urban Westerners have access to many sources of instant gratification, yet these may leave an empty and overstimulated feeling if the recipient lacks intimacy with others, with himself, and with his own body. Such deep states of disconnection may require years of healing before the individual can consistently experience lasting pleasure.

By framing the inclination of organisms towards pleasure as an inclination toward *health*, and by pointing out the ways in which pleasure opens the body's senses to take in and process useful information about the environment, Juhan associates the erotic with human flourishing and self-awareness. As a bodyworker, Juhan sees massage and other bodywork modalities as methods for promoting relaxation and balance. In this state, the body's natural reactions to pleasurable stimuli lead an individual toward a state of holistic health—one that includes a vital internal life and healthy social relationships. He additionally suggests that, by creating a safe space in which to learn to recognize appropriate, pleasurable touch, bodywork may provide opportunities to work through the confusion produced by abuse within emotionally intimate relationships.[31]

In the creation myth discussed above, God Herself comes to self-knowledge by gazing into a mirror, then reaching out erotically to hir reflection in a way that leads to pleasure and self-knowledge. In bodywork, the healing eroticism of therapeutic touch provides a container in which the boundaries between self and other, pleasure and violation, can be examined and solidified. Once again, individuals are "divided for love's sake"—not just for the opportunity to know and love others, but also that they may clearly know their own needs and desires and so love and nurture themselves. Part of knowing where one individual ends and another begins is being able to say which interactions feel loving and connected, and which feel painful and violating. The erotic, says Juhan, contributes to both individual survival and survival of the species by allowing us to make this distinction. The experience of pleasure allows us to know whether or not we are thriving. Just as Lorde claims that being in touch with the erotic makes one aware of alternatives to powerlessness and despair, Juhan suggests that an individual cannot flourish if she cannot distinguish help from harm, kindness from exploitation.

Pleasure as a Sign of Right Relationship

To engage daily with the erotic in a conscious fashion puts us better in touch with our individual human needs (and therefore our health), but it also aligns us with the flow of the universe's ongoing creation. The benefits of accepting erotic pleasure are utterly pragmatic: reduced stress, improved immune response, decreased anxiety, elevated mood, heightened body awareness. Yet erotic pleasure also opens us to meaning drawn in its broadest strokes: our place in the universe, our relationship with the divine, our paths toward joy, ecstasy, and loving connection with others. In Doreen Valiente's

"The Charge of the Goddess," a central piece of contemporary Pagan liturgy, the Goddess declares that "all acts of love and pleasure are my rituals" and "my Law is Love unto all beings."[32] These statements point at something deeper than license for selfish indulgence; rather, they suggest that pleasure is evidence of right relationship. Theologian Marvin Ellison echoes the starry origin myth of God Herself when he proclaims our obligation to align with the erotic power that flows through "sweaty, quivering flesh [. . .] and yet also, magnificently, turns the galaxies and rotates the stars above."[33] As above, so below: the erotic connects us with our inmost hidden selves and also with all of creation, with our loved ones and also with the human species.

From this cosmological understanding of the erotic, an ethic arises in which beauty and pleasure are among our highest priorities. An erotic cosmology takes us deeply into the mystical, into an ego-blurring place of knowing ourselves in God and God in ourselves. Treating the story of the Star Goddess as myth—a sacred story that helps to make meaning in human life—we gain the opportunity to fully and mindfully embrace the body's holy delights. Yet these theological explorations have practical implications. As the following chapters explore, to use the erotic as an organizing principle for an ethical system takes the physical realities of the human body fully into account, and this approach has profound social, religious, and political implications. If we allow the erotic to flow through ordinary touch, we may find that it demands justice: the formation of a society that encourages social and economic equality, values diversity and freedom of expression, and supports the health of individuals and groups.

Chapter Summary

Chapter 1 presents an erotic cosmology grounded in contemporary Pagan myth and liturgy. The erotic is the mutual, pleasurable, fully embodied movement of energy within one or between two or more beings. Although sexual contact can and should be erotic, not all erotic contact is sexual; the erotic is a broader category than sexuality. On a human level, it is a force that urges us toward intimacy, wholeness, and connection; on a cosmic level, it is the binding force that structures Being itself, a primal force of divine creation that is loving, desirous, and carnal. In the creation myth told by Starhawk in *The Spiral Dance*, the universe comes into being through an act of divine lovemaking in which a multiply-gendered deity divides her/him/itself in two (and, perhaps, in the process becomes self-aware for the first time). This story is read as both queer and transgressive: queer in that it sees erotic connection as possible between any and all beings, without regard for gender or even species; and transgressive in that it shamelessly holds out the possibility of erotic connection to all beings, unconditionally. Pleasure is framed as a human birthright, offered as a gift from a universe that is "divided for love's sake, for the chance of union."

The writings of bodyworker Deane Juhan provide a concrete, physical way to approach this erotic cosmology on a personal and social level. Juhan characterizes the existence of pleasure as a successful evolutionary strategy: pleasure is the result when organisms behave in ways that are associated with health and thriving. Though our ability to seek health through pleasurable activities can be undermined by cultural factors that confuse or restrict our sense of what is pleasurable, bodywork and other body awareness techniques can help individuals to cultivate connection with the erotic as a guiding force. Becoming aware of the flow of the erotic in our lives can direct us not just toward improved individual health, but also toward a just and pleasure-valuing society.

CHAPTER 2

Politics in the Pit of the Stomach

PRIMAL PLEASURES AND THE HEALTH OF THE SOCIAL BODY

> *The problems of the individual and the society are no more separable than are those of the body and the mind. [. . .] This is why bodywork put me through lasting personal changes and why it appealed to me as a career: Not because it avoided painful social or political problems for which I had no solution, but precisely because it plunged me into the heart of the matter, into the blood and guts of the issues, turned me inside out and exposed the politics in the pit of my stomach.*
>
> Deane Juhan, *Touched by the Goddess*, p. 15

Before I entered massage school, I was firmly of the opinion that the mind (or spirit, if you prefer) and body were inseparable. I thought that the pervasive Western separation between the two was an arbitrary dichotomy that had outlived its usefulness, if it had ever been useful at all. My belief in the unity of the physical and the spiritual, however, lacked emotional conviction until I stood in front of a class while an instructor read my life history from my body. Although most massage therapists (at least among themselves) speak of "reading the energy" of others almost as frequently as Pagans do, here the basis of the reading was unmistakably grounded in the physical. My slightly collapsed chest and forward head posture suggest the years I have spent typing at a computer and poring over books, but my rounded shoulders also echo a history of depression, my attempt to shelter my heart. My walk and the easy sway of my hips tell of dance lessons, while an uneven standing posture and turned-out feet show my body's attempt to adjust to an old injury. My willingness to make eye contact, the set of my mouth, the tilt of my head—all the experiences of my life have come together in the way I hold my body. For the first time, I realized fully that energy—or spirit—is in no way separate from the body itself.[1] What I, as a practitioner of Paganism, often spoke of as others' "energy" was my intuition's attempt to process incredibly complex information about their *physicality* and present it in a way my conscious mind could use. Yet my massage school crash course in the many possible causes of postural deviations showed me only the tip of the iceberg of what I am able to perceive. My highly social mammalian brain processes an enormous number of details in every interaction with another person, making judgments of which I am only half-aware.

In my career as a massage therapist, I have observed the impact of our society's social and economic patterns on my clients' bodies. The head held too far forward and the collapsed chest are nearly ubiquitous, indicating the frequency of forward-facing, arms-extended activities such as driving and computer use; the same tendency to hunch often accompanies anxiety and depression, which are also common. For experienced bodyworkers such as Deane Juhan, the practice of bodywork brings one face-to-face with the full range of society's ills on a daily basis. He writes:

> The dysfunctional habits and restrictions of movements, the painful symptoms, the pathologies, and the patterns of compensation that I encounter [in the treatment room] are nothing but the physical extensions of my clients' mental states, and these in turn reflect to a high degree their experiences of the social context in which they find themselves. [. . .] Whatever I have avoided in the headlines I am forced to read in the tissues, and in fact it is here that I most inescapably encounter the crux of many global problems. How can an industrialist understand what his factory is doing to the Mississippi River if he can't even perceive what he is doing to his own bloodstream? What can greater social freedom mean to someone who is imprisoned by their own habits? What is patience to someone who is struggling for their next breath, or flexible negotiations to someone who can't bend over, or tolerance to someone with raw nerves, or accountability to someone who will not take personal responsibility for the condition of their own body?[2]

Western society largely ignores Juhan's obvious insight: groups are made of individuals, and individuals' health has a profound effect on group behavior. A society of individuals in chronic yet seemingly meaningless low-level pain is unlikely to foster consistent social attitudes of flexibility, tolerance, or altruism. We all know the effect that a persistent headache has on our ability to be our best, most polite, most compassionate selves in the workplace or with our families. What effects might our collective levels of stress and touch deprivation have on our social dynamics? What might shift in a society less influenced by the Protestant work ethic, where workers (due to economic necessity, company policy, or an internalized fear of feeling lazy) do not feel obligated to push through illness rather than staying home to recuperate properly? Would our voting patterns change if, every day, each one of us was able to routinely attend to the needs of the body—good food, sufficient exercise, physical affection from loved ones, restful sleep? Juhan continues:

> [Bodywork] taught me that my active engagement with my body was the source of any and all solutions that might be brought to bear on my situation—not because my body was the locus of all problems, but because it was my only means of perceiving, understanding, and dealing with any of them. Because it showed me that my body is my personal world, and that the qualities of that world are spun out of the consciousness of my body.[3]

Juhan writes that the body is the only lens through which the individual can "perceive, understand, or deal with" life's problems—the body is the foundation of consciousness and determines its characteristics. Underlying Juhan's assertion is a belief in the full integration of body and spirit, a belief Juhan shares with many contemporary Pagans. From this perspective, bodily ills and spiritual ills are inseparable. The failure to value the pleasures that support the body's health and well-being, therefore, necessarily contributes to a less just, less efficient, and less compassionate society.

The Erotic as Moral Force

The relationship between individuals' access to pleasure and the overall health of society is a major concern for Christian theologian Marvin M. Ellison. In *Erotic Justice*, he claims that the failure to value erotic pleasure is one of the cornerstones of social and economic injustice in Western society. "Every oppression involves violence toward the devalued body," he writes,[4] whether that oppression involves domestic abuse or sexual assault, mandatory unpaid overtime or sweatshop working conditions, laws criminalizing consensual sex acts between adults or hate crimes committed against minorities.

Ellison chastises the pre-feminist, pre-gay rights liberal theological tradition in which he was educated on two counts. First, he criticizes its failure to fully recognize the importance of power inequalities when working for social justice. Second, he attacks its inability to recognize erotic love as a divine power with significant moral force. Liberal theology, says Ellison, has followed Western culture as a whole in splitting life into public and private realms. Economic and political relations are seen as public, while the family and sexuality are private. Regarding the public realm, liberal theology speaks of social justice as a practice, with the ultimate goal of a society that respects human dignity and encourages individual freedom and well-being. Love, however, is restricted to the private realm of family, sexuality, and romance. In its separation from public life, love becomes "politically ineffectual and reduced to an affective sentiment"[5]—a sweet emotional state, not a force that pushes one to compassionate action.

This public/private split, says Ellison, obscures the larger social impact of power inequalities in the "private" realm. Further, it discourages individuals from applying the tender and connected ways of "private" relating to public relationships.

> Not only does liberalism fail to associate intimacy with justice, it also bypasses the deficits of love without justice. The romanticizing of marriage and idealizing of family life conceal widespread abuse and human suffering. The notion that family life is private blocks recognition of how the quality of personal life is dependent on the wider social order.[6]

Ellison believes that the conditions of society and those of the personal lives of individuals cannot be meaningfully separated. For example, even in our supposedly post-feminist era, the fact that women's earning power as a group remains considerably less than men's puts many women at an economic disadvantage in romantic relationships. This inequality has a direct impact on women's romantic and sexual choices, particularly when children are involved.

> [M]ost men, especially white affluent men, feel entitled to exit any relationship when they choose, and they have the power and resources available to do so. The freedom to extricate oneself from an undesired situation, however, is denied to all but the most economically privileged women. Most women lack the independent economic resources that would allow them to break free from abusive situations and leave loveless marriages.[7]

Romantic relationships are strongly affected by larger social inequalities, says Ellison. An uncritical admiration of the institutions of marriage or the conventions of romantic love can miss the ways in which such practices can contribute to systematic violations of individuals' bodies, both small and large. A woman who allows her partner to beat her because she does not have the social or economic resources to flee is being wronged, but so is a woman who remains married and agrees to sex she does not want because she cannot otherwise support her children. Both live lives in which the autonomy and pleasure that is every individual's birthright is denied.

Seeing love, and especially erotic love, as apolitical prevents us from applying the insights that come from connected intimate relationships to the "public" realm. A joyful sexual connection, says Ellison, flows between individuals who mutually respect and affirm each other's right to have pleasure and to govern their own bodies. In his view, to affirm these rights strikes at our society's tendency to eroticize power inequalities. It also challenges the implicit belief that those in power rightfully control the behaviors and exploit the bodies of the politically disenfranchised, whether by law or through custom.[8] The attitudes that lead to mutually pleasurable and sustainable sexual relationships have profound implications when applied politically, economically, and socially. Erotic love need not be a mere sentiment, but can rather be a force that urgently advocates for the physical health, pleasure, and well-being of individuals and groups.

Ellison particularly notes that in American culture, black sexuality has been stereotyped as dangerous and chaotic, and that this characterization has been used to justify exploitation of black bodies. He believes that such prejudices play an important role in African-American subjugation.[9] The United States continues to experience the aftereffects of its heritage of slavery, under which white slaveholders literally owned the bodies of their slaves and slave women were used as concubines and wet nurses. The slave-era belief that black persons are more animal than human contributes even today to destructive mischaracterizations of black sexuality. Ellison writes:

> When women of color are raped, they are seen as naturally promiscuous, less worthy of protection, and deserving what they get. [. . .] In addition, men of color are frequently scapegoats for other men's sexual violence. Among incarcerated rapists, a disproportionately high number are men of color convicted of sexual assaults. Among these, a high number receive the most severe sentences, including capital punishment. [. . .Yet r]ape is a crime occurring most frequently *within* social groupings and, statistically, remains predominantly a white male crime against white women.[10]

Racist assumptions serve as linchpins for the exploitation of individuals' bodies: exploitation that denies disempowered individuals the right to pleasure but bolsters the social and political status of those in power. Ellison also sees Western society's construction of heterosexuality as contributing to this problem, because dominating women or less powerful men is part of how Western culture has understood what it is to be a "real man."[11] Although gender roles are becoming more fluid in contemporary society, we still hear of men who are sensitive, affectionate with other men, or unassertive referred to derogatorily as "sissies" or assumed to be gay (and, therefore, "not real men"). Assertive women who take on leadership roles are still sometimes accused of being unfeminine, potentially lesbian "ballbreakers" out to usurp men's rightful place of dominance. Such judgments on how well individuals conform to normative gender roles, says Ellison, "pollute the channels of sexual intimacy on which people depend for open and trustworthy communication."[12] Erotic energy cannot flow when men feel pressured (by the expectations of others, and perhaps by their own internalized expectations) to dominate and control those around them. To fix individuals into dominant or submissive power roles on the basis of gender or sexual orientation presses them into stereotypical molds that restrict the genuine self-expression and honesty necessary for intimate relationships.

The Erotics of Power

Ellison claims that in order to combat attitudes about sexuality that contribute to oppression, advocacy of erotic justice "*requires the eroticizing of equality between persons and among groups.*"[13] Ellison does not address consensual sadomasochistic power exchange, though he admiringly cites Gayle Rubin's essay "Thinking Sex," which defends consensual adult sexual behavior in general and BDSM in particular (see my discussion in Chapter 3). One might initially think that because BDSM practitioners regularly play with the roles of dominance and submission, their values would be in conflict with the idea that equality should be eroticized. The BDSM community has a strong emphasis on negotiation and consent, however, in which the value of equality is implicit. Additionally, the conscious exchange of power between people actually calls attention to power inequalities in daily life, as well as providing a vocabulary with which to analyze them; BDSM practitioners are apt to recognize and discuss unexamined D/s (dominance/submission) patterns in their workplaces, neighborhoods, and social circles. The unique perspective of the kink community has much to contribute to the erotic freedom and health of all people, including those whose desires are thoroughly mainstream.

In her writings, including the books *Indecent Theology* and *The Queer God*, postmodern Christian theologian Marcella Althaus-Reid argues that those who are usually considered sexual "deviants" offer a uniquely liberating perspective on sexual issues.[14] Althaus-Reid comes out of the liberation theology tradition, which emphasizes political and social liberation from oppression and frequently speaks of a "preferential option for the poor." By this, theologians mean that in thinking about social justice, the perspectives of the poor and oppressed are considered especially valuable. Because economically underprivileged people are excluded from the systems of power that make up the status quo, their point of view is prophetic,[15] potentially infused with divine presence. Of all those taking part in the conversation about social justice, those who have nothing left to lose but their lives are in a special position to speak about injustice. In liberation theology, this position has the potential for remarkable clarity. The oppressed can offer insights to those with more power (such as educated theologians) who see that existing systems are unjust, but feel implicated and restricted by their own privileged positions in them.

Althaus-Reid's claim for sexual minorities is that because their erotic expression is marginalized and feared by mainstream society, the risks they take to seek authenticity, pleasure, and even ecstasy open the door to powerful and unusual experiences of divinity. In her view, the divine cannot be encompassed by tradition or institution; when institutions resist change and become rigid, they cut off their access to God's presence. Understanding the queer perspective as fundamentally subversive, Althaus-Reid sees those on the margins as key voices in the conversation about sexuality. Their views may challenge conventional wisdom, but their contributions speed change and help to prevent spiritual stagnation.

As discussed in the Introduction, healthy practitioners of BDSM approach power exchange negotiations with the attitude that both parties begin the negotiation on equal ground. Before taking on the roles of top and bottom, individuals come to agreement on erotic activities that they both enjoy. Only once the scene has begun do the players allow their consciousnesses to shift, with tops embracing the thrill of control while bottoms experience the exhilaration of letting go. At all times, both top and bottom can use a specially chosen "safeword" to end the scene. A successful encounter must satisfy the desires of both parties. A top who fails to create a safe emotional container for intense experiences or who does not respect her bottom's wishes will often lose her playmate at the end of the session. Additionally, in urban areas where active BDSM communities gather, community accountability helps to ensure ethical behavior. Obvious abuses of trust will damage players' reputations within the community, and make it difficult for them to find willing partners.

Obviously, in normal society there is no way to call "safeword" and end the sexual harassment of an authority figure, the homophobic threat of violence from fellow students or co-workers, or the experience of humiliation by structures of institutionalized racism. "Patriarchal sex is alienated sex," writes Ellison, asserting his belief that when one group has the power to systematically control, degrade, and exploit another, true erotic connection between individuals from the two groups becomes difficult or impossible.[16] Yet the existence of the BDSM community, in which men, women, and transgender people explore roles of dominance and submission in every combination of race and gender imaginable, suggests that to find positions of unequal power sexually arousing is not unusual, nor necessarily pathological. Practitioners within the BDSM community commonly speak of the trust and intimacy that can be developed through partnerships in which negotiated power exchange takes place. Kinky sex differs from what Ellison calls "patriarchal sex" because it aims to connect, not alienate. In such bedroom games, neither individual is deprived of or given power because of their gender, race, or socioeconomic class. Rather, power is fluid, temporarily given away or accepted by choice, with the intention to maximize pleasure (or erotic pain) and excitement for both parties. This is not an ethic of control, but one of experimentation and exploration. In its norms about negotiation and its assertion that all participants have an equal right to pleasure, the kink community shares Ellison's values of equality and respect.

Additionally, Ellison advocates for understanding the erotic as a mode of communication, one necessary for moral knowing. Similar to Juhan's claim that the body is the only lens through which the challenges of life can be understood and dealt with, Ellison states:

> As sensuous human beings, we know and value the world and therefore become self-directing moral agents only as we feel connected in and through our bodies. As Beverly Harrison contends, "*all* our relations to others—to God, to neighbor, to cosmos—[are] mediated through our bodies, which are the locus of our perception and knowledge of the world." [. . .] Moral knowing is rooted in feeling, and we depend upon sensuality to grasp and value the world. When sexuality is feared and evaded, people lack responsiveness and run the risk of becoming out of touch with what causes joy, suffering, and vulnerability, including their own. A people alienated from their bodies are more likely to be content with, and even at home with, pain and oppression.[17]

Ellison also echoes Audre Lorde's statement that being in touch with the erotic makes oppression less tolerable, its wrongness clearly revealed by the body's displeasure. Again, displeasure is not the temporary discomfort of hard work or an ultimately healing medical procedure, but rather the draining, long-term pain of illness, depression, degradation, or ill use. "Patriarchal Christianity has it wrong," writes Ellison.

> The erotic is not a hostile, alien force lurking from within to bring us to ruin, but is rather an internal moral guidance system, grounded in our body's responsiveness to respectful, loving touch. [. . .E]rotic desires are not inherently selfish or antithetical to moral value. Progressive seekers of justice-love can well imagine living by an ethical eroticism that enjoys life's pleasures and at the same time prods us to pursue a more ethical world. The erotic can fuel our passion for justice. [. . .] Defenders of the status quo rightly see erotically empowered people as dangerous and beyond their control.[18]

Like Lorde, Ellison sees the erotic as a divine force that presses for liberation, a swelling stream that threatens to burst the dams maintaining the power relations of mainstream society. He does not argue, however, that erotic urges are inherently moral; all our desires are shaped by our culture and can be channeled destructively. Instead, he lays out a set of ethical values that characterize a grounded, connected eroticism.

For Ellison, ethical eroticism has four central values. First, the body's goodness, as well as eroticism and sexuality, must be honored. Second, the body's integrity and the right of the individual to authority over her body must be guaranteed. Because there is no separation between mind and body, respecting the dignity and autonomy of a person must include respecting her physical being. Third, ethical eroticism is mutual and consensual, aiming to produce good communication and a sense of connection. Finally, fidelity—the keeping of commitments and the ability to be honest about one's needs and desires—provides a necessary container for erotic relationships.[19] From these principles, says Ellison, it is possible to develop an ethic that eroticizes justice on a social level, not just on the level of personal relationships. He envisions a society in which many forms of relationship, not just heterosexual marriage, are acknowledged and celebrated, and he mentions the potential for ethically nonmonogamous relationships.[20] In addition to advocating for responsible sex education for adolescents and for empowering people to protect themselves from unwanted touch (as well as for holding perpetrators of sexual violence accountable), Ellison asserts the broader principle that "[a]n ethic of erotic justice *from the start rules out relations in which persons are abused, exploited, and violated*."[21] Such an ethic potentially addresses issues of economic and social inequality that have nothing directly to do with sexuality, gender, or sexual orientation. Defending the individual's right to erotic pleasure—connected experiences of consensual, pleasurable touch—becomes the center of Ellison's sense of what constitutes ethical living.

Accepting the Primal

To my ear, however, there is something thin and unconvincing in Ellison's claim that the key shift Western society must make in its sexual culture is to eroticize equality. The dangers of fixed hierarchies are clear—our society is riven by systematic structures of inequality, and these inequalities clearly can poison romantic relationships in the ways Ellison describes. Yet his argument does not address the play of the BDSM community, where gender fluidity is common and women sometimes top men and other women. As I will discuss in more depth in Chapter 3, however, some feminists have attacked the kink community as simply imitating Western society's patriarchal structures, regardless of the way it might empower individual women. Women's participation in power exchange is seen as wish-fulfillment that does not address broader issues of gender inequality in society.

One such writer is Goddess feminist Riane Eisler. Although Eisler does not speak from the Christian tradition, her beliefs are based on a story of a fall from grace, much like the story of Adam and Eve upon which the Christian doctrine of original sin is based. In her books *The Chalice and the Blade* and *Sacred Pleasure*, Eisler embraces the feminist myth that in our prehistoric past, the people of Old Europe lived pacifistically in an egalitarian, Goddess-worshipping society.[22] According to mid-twentieth-century archaeologist Marija Gimbutas, this peaceful civilization was conquered by a patriarchal, warlike Indo-European tribe that invaded from the east between 4500 and 3000 BCE. Eisler and other feminist writers trace social hierarchy between the genders and a wide range of violent behaviors to this supposed cultural shift. In Eisler's view, BDSM is a symptom of the subsequent millennia of patriarchal domination, and such behavior will cease once our society again embraces an egalitarian, partnership model of relationship. This narrative of matriarchal prehistory has been questioned by modern feminist scholars, however;[23] while polytheistic, relatively egalitarian Neolithic societies may have existed, the pacifistic Goddess-worshipping utopia envisioned by Eisler is at best an exaggeration of a historical reality. Despite the weakening of the matriarchal myth, the belief that it is not natural to enjoy physical aggression, bondage, or dominant/submissive behaviors lingers in feminist circles.

Rather than ascribing some human beings' enjoyment of violence to the long-dead Indo-Europeans, I see the capacity for dominant and submissive behaviors as part of our animal biology. The two primate species to which human beings are most closely related engage in a range of hierarchical behaviors. Chimpanzees tend to be hierarchical and frequently violent, while bonobos (pygmy chimps) demonstrate less territoriality and are more oriented toward sharing and group bonding.[24] Anthropologists have suggested that by studying the social behaviors of these two species, we may be able to better understand our own social capacities as human beings. In fact, popular coverage of bonobo studies has helped to weaken public perceptions that "alpha male"-based societies are universal to human origins. Human beings are capable of a range of hierarchical behaviors but, as in our cousin species, such urges do not express themselves with equal strength in every individual or within every group. Additionally, culture mediates our biology in a variety of ways, some of which are clearly destructive, while others may be empowering and liberating.

This nuanced perspective has no place in an ideology that looks back to a mythical period of perfect egalitarianism. Eisler categorically dismisses the experiences of feminists who enjoy BDSM, writing:

> to equate sexual freedom with sexual excitement from the chaining, torturing, humiliation, and degradation of others seems totally insane. And in the sense of insanity as a failure to perceive reality, it is. But in the context of a socialization to equate sex and violence, to mistake pain for pleasure, and to experience "loving" as hurting or being hurt, it is understandable. And the fact that this socialization remains largely unconscious makes it even more powerful.[25]

Eisler can only deal with the statements of feminist BDSM practitioners by calling them insane, characterizing their play as "torture," and claiming that they are "mistaking" pain for pleasure. This patronizing stance dismisses the experiences of ecstasy and connection that practitioners of BDSM describe. Experiences of deep trust and vulnerability, as well as exploring the boundaries of their physical limits, are common reasons that practitioners give for engaging in activities involving power exchange.

In their books on BDSM (*The New Bottoming Book*, *The New Topping Book*, and *Radical Ecstasy*), Dossie Easton and Janet Hardy discuss the way that BDSM can harness aggressive, primal desires in a way that *resembles* acts of violence, but is not *experienced* as violence by the participants. In fact, according to Easton (whose primary employment is in counseling), such play can actually be part of a process of healing from abuse or assault.[26] Clearly, this kind of deep emotional exploration is risky, and is perhaps best reserved for those who have already healed as best they can in a traditional psychotherapeutic setting. Some assault survivors, however, find healing in BDSM play with trusted loved ones because the role-playing allows them to confront abuse-related emotional triggers in an environment that is under their control. Far from being necessarily violent, BDSM may actually provide opportunities to heal from violence.

Gayle Rubin remarks that BDSM is, physically speaking, considerably safer than football,[27] which routinely leaves professional players with broken bones and brain damage from repeated concussions.[28] Certainly I would be traumatized by the experience of being chased, tackled, and piled upon by a group of football players, even ones approximately my size and shape, yet football players both pro and amateur find the game exhilarating. Commentators have frequently remarked that sports, particularly those that lend themselves to rough interactions, may channel human urges to fight and even make war into enjoyable competitions that challenge players emotionally and physically. While there are certainly many people who do not find the idea of being tied up and flogged—or tying up and flogging someone else—to be appealing, to me this is a matter of taste similar to liking or disliking contact sports. In fact, for the average BDSM practitioner, a taste for sensation play may be better compared to a liking for very firm deep tissue massage. Most BDSM play leaves the recipient with little more than a reddened bottom, a few welts or bruises, or some strategically placed bite marks. With the risk of broken bones or concussions, contact sports are much too extreme a comparison to the activities of all but the most adventurous and experienced BDSM practitioners. Accordingly, I see BDSM as a relatively safe and controlled way that practitioners connect with primal instincts. Instead of hunting and killing for food, struggling for territory, or making war on our neighbors the way our ancestors did (and other primate species still do), BDSM practitioners have found ways to deepen intimacy and experience exhilaration by harnessing these urges.

Hierarchical power relations are not inherently pathological; in fact, dynamically exchanged power provides opportunities for connection. When one person temporarily hands over responsibility for her physical and emotional well-being to another, states of profound vulnerability and honesty become possible. BDSM practitioners are also able to acknowledge that human sexuality, when fully unleashed, is rarely civilized or conventionally pretty. Sexuality connects us to our animal nature; even within mainstream media, we recognize images of biting, scratching, or crying out as signs of particularly unrestrained, enthusiastic lovemaking. Many of us have probably nursed a sore muscle or a bruised knee on the morning after an exciting sexual encounter and smiled at the memory of a primal pleasure.

Historian of contemporary Paganism Chas S. Clifton sees the notion of "the wild" as being fundamental to Pagan theologies of the body, especially in the United States. In Pagan culture, anti-authoritarian politics are linked with a connection to the natural world and to control over one's own body. Drawing on the ecopoetics of Gary Snyder, Clifton associates Pagan political and spiritual values with Snyder's understanding of the wild as being contained within the human: qualities of being spontaneous, fierce, and self-reliant, unhindered by cultural conditioning.[29] These values are implicitly dramatized within many Pagan rituals. Some traditions employ sensual elements to invoke a kind of primal timelessness: rituals may include anointing with oil, drumming, chanting, incense, exchanging kisses around a circle, and/or dancing (with or without bonfires). Many Pagans honor deities who personify the wild: spirits of the hunt, deities of intoxication and ecstasy, or guardians and caretakers of beasts. Many Wiccans practice ritual nudity as an affirmation of the body's sacredness and a demonstration of the participants' freedom and intimacy with the natural world. By sacralizing the body and sexuality, argues Clifton, Wiccans and other Pagans effectively framed a "wild" aspect of human life in a nature-based religious context.[30] For many Pagans, therefore, connecting with wild nature begins with turning to one's own body.

Ellison's argument lacks this acknowledgment of the wild messiness of sexuality and fails to acknowledge a certain amount of aggressiveness—on the part of both women and men—as normal to some erotic connections. His desire to eroticize equality and respect is admirable, but it does not address the possibility that certain hierarchical tendencies may be inherent in some human animals, just as they appear in our primate cousins. We do not listen to the yowling of alley cats forcefully copulating in an alley and call it rape. Although it is true that as animals increase in complexity, issues of consent arise—some nonhuman primates have sophisticated enough communication to consent to sex or refuse it, and so sexual violation becomes possible, just as it is for human men and women[31]—the fact that we are animals who can express our complex desires does not strip us of our fundamentally animal nature. No one would reasonably criticize a dog for its servile behavior towards its caretaker, or consider the pecking order of a dog pack to be, intrinsically, an unethical violation of the bottom dog's rights.

Our far greater self-consciousness and more complex ethical nature gives us more choices than dogs about where we allow hierarchies to arise, how rigid they will be, and how such structures can serve all those within them. But our desire to moderate the hierarchical tendencies in our animal makeup will not eliminate them. It is normal, not pathological, to at times enjoy being taken care of or following orders; it is also normal to at times enjoy caretaking or giving the orders. As self-conscious animals, however, we have the awareness that these are roles, and they can be cast off at any time. Therefore, whether male or female, we need not fear that our status as feminists, egalitarians, or moral persons is threatened because we enjoy being tied up, spanked, bitten, dressed in clothes of the opposite gender, or called "baby." If connection and empathy are present and power is able to flow freely in our erotic relationships, such roles can become a source of pleasure and self-exploration rather than restriction. Experimentation that happens in the bedroom can expand our awareness of what is possible—each individual's potential to command or yield, chase or surrender—and to bring that awareness of our changeability into other social relationships. If a person can command in the bedroom, does it open up the possibility that he might also take on unaccustomed leadership roles at work or at his place of worship? Might a person who suffers because she cannot control every aspect of her environment benefit from having a trusted loved one deprive her of control for twenty minutes and have that experience be a pleasurable one? Certainly, as with every type of human interaction, BDSM play can be engaged in irresponsibly and abusively, but BDSM also presents unique opportunities to explore one's self and relationships in a way that challenges the status quo.

Empathy as an Answer to Inequality

Ellison's principle of "mutuality" is a more compelling central principle for erotic ethics than equality. Ellison wishes to eroticize equality so that the pleasure some people take in nonconsensual domination will be neutralized. Yet this stance misses the root of the problem, which is a lack of empathy. Empathy as defined in research psychology is the ability to share the perceived emotion of another person. It is an emotional response triggered by relating to another person that resonates with the other's perceived state. Empathy differs from sympathy in that it does not necessarily involve the impulse to change the other's emotional state, for example, to relieve their suffering. It is also distinct from emotional projection in both direction and quality. One may project one's attitudes onto another person without including an emotional component, and the direction of the projection flows from self to other. In empathy, the direction of flow is from other to self, and the response is necessarily emotional.[32] Because empathy training has been found to increase group cohesion, improve individuals' social skills, and encourage tolerance and teamwork within diverse groups, programs designed to increase empathy in young people are now being implemented in some schools.[33] Several recent books have also drawn on empathy research to suggest that, in a time of dwindling natural resources, empathy practiced as a guiding social principle could lead to more sustainable social, political, and economic structures.[34]

Because perfect economic, social, and political equality between all partners is impossible—and Ellison himself argues strongly for the importance of respecting diversity within sexual relationships—eroticizing the ability to relate to and care about another's experience may better address the harms of patriarchal culture. When we eroticize the relational capacity, the practices of the kink community come smoothly under the umbrella of ethical eroticism, as do the mildly kinky bedroom activities of millions of more mainstream Americans. A woman who enjoys being tied up and a man who enjoys tying her may appear to be playing out unhealthy heterosexist dynamics of male dominance and female submission, male sadism and female masochism. Yet when that male partner is attendant to and invested in his female partner's pleasure and vice versa, their interaction becomes an example of erotic connection and mutuality, not of alienation and exploitation. Empathy, rather than equality, may be a more reliable starting point for evaluating the ethics of a given relationship.

There is a utopianism to liberal Christian theology (including Ellison's) that I have always admired. Particularly since the "Social Gospel" of nineteenth-century Christianity, many liberal Christians have held a vision of "the Kingdom of God on earth" where social ills are cured and people live together in relative harmony. Liberal Christian faith drove a variety of nineteenth- and twentieth-century reform movements, from the passage of labor laws preventing the exploitation of workers, to women's suffrage, to the civil rights movement of the 1960s—just as it is an active force in the gay rights movement today. This underlying utopianism lends a sense of sacred mission, buoying reformers up as they engage the messiness of the political system and confront those who oppose change.

I have mixed feelings, however, when theologians imply that the darker urges of human beings can be eliminated. Particularly problematic is the Christian doctrine of original sin, the belief that human beings are born wrongdoers and need outside salvation in order to avoid ultimate damnation. Yet original sin is not the foundation of all Christian belief. There is a persistent, contrasting thread of Christian theology (recently popularized by Matthew Fox[35]) that humanity is made in God's own image and so reflects an *original blessing*. Human beings are not born as sinners, but instead contain equal potential for good and evil. Building on this belief, liberation theologians like Ellison (as well as feminist Pagan and Goddess theologians like Starhawk and Eisler) locate human evil largely in the social realm. There, flawed, hierarchical systems of relationship, having risen for historical reasons, must be confronted and replaced with non-hierarchical, egalitarian alternatives. Both Eisler and Starhawk imagine egalitarian, communal, pluralistic societies,[36] and their visions fit Ellison's ethics well.

Yet Ellison's insistence on *only* eroticizing equality is problematic, because it threatens to deprive us of something fundamental to being a human animal. Certainly encouraging others and ourselves to find peer relationships sexy is a good thing; as previously mentioned, BDSM negotiations about power exchange must take place between functional peers in order for consent to be valid. The ultimate container for an ethical BDSM relationship is a peer relationship, to which both parties are returned once a play session has ended or the safeword has been called.[37] Naturally, it can be difficult to determine whether a peer relationship has been fully established, due to the social and economic inequalities that are systemic in our culture. One might object that a woman of small economic means does not have an equal negotiating position with a wealthy man, and so her consent to be dominated is always suspect. But how does this situation differ from mainstream heterosexual marriage, in which women with little money often agree to binding legal contracts with wealthier men? Some radical feminist analyses suggest that in such a situation, the consent of the person with less economic and social power is subject to pressures—namely, need for stability and safety—that make his or her consent suspect. Yet to invalidate an adult's ability to give consent, to choose between yes or no, dehumanizes and disrespects them.

If the feminist movement began with the belief that all people should be afforded the same opportunities regardless of gender, then to tell a woman that she does not have the ability to choose whom to marry or what kind of sex to have betrays that movement utterly. We can speak of the pressures that compromise the ability to consent, and of social reforms that would reduce such pressures, without denying adults their independent agency. While eroticizing equality, as Ellison says, we must also respect *diversity*.[38] Ellison speaks of honoring sexual diversity: affirming a variety of relationship models and sexual styles, as well as the fact that intimacy needs may change over the course of a lifetime (he himself identifies as a gay man who, when younger, lived contentedly as a self-identified heterosexual). Yet I would take the need to honor diversity further: diversity must be respected *within* sexual relationships as well. Even as we strive for peer relationships, we will never all be equally intelligent, wealthy, physically attractive, sexually adventurous, or socially adept. Yet we can still reach toward a world where, despite these differences, people do not routinely enter sexual or romantic relationships in which one person rigidly holds power over another.

Re-Evaluating Hierarchy

Ellison's use of "equality," a key term of the feminist and civil rights movements, means equal treatment under the law and equal opportunities, not homogeneity. Today "equality" is a concept dear to those working for legal same-sex marriages across the nation. Yet Ellison—and Starhawk and Eisler—still see the tendency to enjoy and eroticize hierarchy as a pathological side-effect of history, not a value-neutral part of many human beings' biological makeup. When Ellison says we must eroticize equality, he implies that the darker tendencies of human beings can be obliterated. As in the liberal Christian Social Gospel, Ellison suggests it is possible to be free, not just of violence, but of aggressive urges; to be free not just of economic inequality, but of the desire to dominate.

Other progressive Christian theologians have addressed the question of BDSM directly. Carter Heyward, writing a decade before Ellison, is inclined to see it as an acceptable form of sexual expression, but she implies that it is a symptom of human brokenness, not an aspect of humanity's having been created in God's image. In *Touching Our Strength*, she writes that due to the unhealthily hierarchical nature of our society, we do not learn mutuality as a natural way of being, but instead learn masochism (as she defines it, the desire for punishment) and sadism (the desire to hide our own vulnerabilities by abusing those who are more vulnerable).

> [W]hile our sadomasochism is pathological, the disease is not primarily one of individuals. Sadomasochism belongs to the whole society. [. . .] But can sadomasochistic eroticism be a relational conduit through which we move toward mutuality not only with each other but also with God, the source of our liberation? The answer is that it *must* be, because we can reach one another and God only from where we are here and now. [. . .] "Having sex," if it is erotic, is about power-sharing. As such, it involves journeying together through places of brokenness and pain toward safety and tenderness. Sadomasochistic eroticism does not signal necessarily that something is wrong with us individually, but rather indicates, unmistakably, how fundamentally formed we are—emotionally, spiritually, physically—by the world we inhabit.[39]

Heyward believes that we must begin where we are, in a state of brokenness and pain, and that if we reach out to each other in love, even unhealthy sadomasochistic patterns can open up sacred erotic connections between us. By "sadomasochism" here, Heyward includes explicit BDSM activities such as bondage or roleplaying. Primarily, however, she is addressing normative heterosexual behavior and expectation—the expectation that in intimate relationships, power inequalities, chronic self-sacrifice, and a certain level of emotional abuse is just part of "the realities of marriage" (or partnership). On the specific topic of BDSM, she adds that sadomasachistic cultures have

> infuse[d] many women and men with an embodied confusion of violence and ecstasy. [. . .] Whether sexual s/m is also a *moral* confusion—that is, whether it is always wrong—seems to me a complicated question with no easy or fixed answer. It is wrong physically or emotionally to hurt one another purposely, knowingly to abuse one another's bodies or psyches, whether or not "with mutual consent." But is it wrong to explore together the limits of our capacities for pleasure, or pain? To struggle together in the tensions and pathos of being more or less in control of our lives, dreams, and destinies? Such explorations can take the shape of s/m fantasies and activities. They do not seem to me necessarily wrong, just very deeply human in an honest, poignant, even at times playful way.[40]

Although Ellison is less explicit about his beliefs surrounding BDSM, his focus on eroticizing equality suggests that he shares Heyward's conflation of aggressive impulses and hierarchy with "violence." Yet "violence" is not the language that the BDSM community uses to understand itself. Heyward questions the kink community's moral status, but also demonstrates her familiarity with that community by using its own terms: "exploring the limits of pain and pleasure," "struggling together," "letting go of control." Such terms are consistent with the writings of the community's own representatives. For example, in *The New Bottoming Book*, Dossie Easton and Janet Hardy speak of bottoming as a process of finding one's own strength and undergoing ordeal to explore vulnerability and resilience. Topping involves responsibly channeling powerful emotions and sensations for the benefit of someone who has placed him- or herself in the top's hands. The exploration of both roles, according to Easton and Hardy, leads participants towards an embrace of their own power. They describe the "alchemical bottom," emphasizing the glorious experiences of intensity, confidence, and beauty that BDSM can provide:

> You, as a bottom, are a crucible. You take the top's energy into you. Within you, liquefied with the heat of spirit and sex, that energy swirls together with your own energy, and turns into something stronger, sharper, brighter than either of you could generate alone.
>
> When your top believes you are beautiful, you become beautiful. When you believe your top is powerful, he becomes powerful. Your

> body and spirit are the locus for the creation of a newly minted reality . . . a reality with all the power and wisdom that you and your top can together bring to it.[41]

For Easton and Hardy, BDSM play exemplifies what many feminists call "power-with,"[42] as Starhawk defines it, "social power, the influence we wield among equals."[43] Shared power, for Easton and Hardy, is power that grows; assisting others in experiences of their own power leads everyone involved to feel more agency, not just sexually, but in many areas of life.

In contrast to Easton and Hardy's view, Heyward and Eisler's characterizations of BDSM demonize tendencies and behaviors that are widespread in human beings and are potential sources of beauty, strength, and pleasure. Ellison too sees the source of the problem as hierarchy itself, not rigidity: the fact of inequality, rather than a lack of empathy between persons in different social or economic positions. While I agree with these theologians that social and economic inequality in our society causes suffering and does indeed need reform, Ellison, Heyward, Eisler, and Starhawk consistently condemn hierarchy without considering how it is implemented.

I studied and taught for some years in the Reclaiming tradition of witchcraft, an ecofeminist strand of Paganism of which Starhawk was one of several founders.[44] Embracing the feminist values of the late 1970s, Reclaiming strives for nonhierarchical forms of organization and uses consensus in its decision-making process. In the last decade, however, there has been increasing acknowledgment—as in other communities that have been influenced by activism—that in explicitly nonhierarchical communities, "shadow hierarchies" inevitably arise based on ability and forcefulness of personality. Such hierarchies are difficult to acknowledge when the group as a whole is committed to radical egalitarianism. Almost inevitably, pointing out a shadow hierarchy feels like an attack on those who are putting the most energy into making the group run smoothly and who therefore have the most control. Further, to acknowledge the shadow hierarchy involves recognizing that the group's behavior differs from its values, a move that can feel like admitting failure.

Reclaiming witches are still very much in dialogue about how to share power within community and, when power is invested in a particular individual for the purposes of teaching or organizing, about how to keep power moving through the group dynamically rather than having it rest long-term with a specific leader or leaders. Recognizing the need to respect and acknowledge the abilities of elders in the community while also wishing to help newer members take on leadership roles, Reclaiming witches continue to debate exactly what "nonhierarchy" looks like in practice.

Without outright rejecting the principle of nonhierarchy, large Reclaiming events have often functioned using temporary but fully acknowledged hierarchies, as individuals whose abilities are known step into the roles of teacher and organizer to consciously mentor others. The act of a group investing its power in a few representatives is not in itself problematic; rather, the danger lies in the possibility of that relationship becoming fixed. Recent criticisms by some Reclaiming elders, however, suggest that these efforts have not prevented power hierarchies from becoming entrenched, nor have they prevented forceful personalities and influential alliances from using intimidation and bullying in the consensus decision-making process.[45] Additionally, the tradition has not been successful in consistently supporting its most experienced members, and some dynamic leaders (including tradition co-founder M. Macha NightMare[46]) have left the tradition, either out of frustration, or to pursue advanced training or leadership in other organizations. Because of these failures, some current and former members have challenged the tradition's commitment to nonhierarchy, claiming that at least as currently structured, it is an ineffective method of ensuring power-sharing.[47]

I reject the idea that hierarchy inherently causes human suffering. As self-conscious human beings, we have some power to choose when and how we engage in dominance or submission. Rather than seeing these as learned behaviors that would disappear in people who live in a wholly egalitarian society, I see both roles as inherent in all people—qualities that can be used to increase connection and intimacy or interfere with them. Clearly, the relative rigidity of social and economic hierarchies in our society and in some of our relationships are a source of pain, but in the BDSM community—where power dynamics are named and negotiated openly—they become a source of pleasure and a mechanism through which players gain knowledge about themselves.

Rather than seeing the BDSM community as deviant, I see its practices as part of a spectrum of human erotic behavior, milder forms of which (scratching, light bondage, spanking) are common in more mainstream sexual relationships. Although explicit BDSM activities may simply not be right for some people—either because their erotic tastes are gentler, or because of abuse histories—the urge to behave aggressively towards a loved one is not necessarily a violent one. The lack of consent and violation of boundaries that makes violence emotionally damaging need not be present in an aggressive erotic exchange. Instead, an empathic connection in which both parties remain concerned with and aware of the other's desires and emotional state is possible. In such a relationship, human beings' most primal, physical urges can be explored for pleasure and mutual growth.

Nature and the Problem of Evil

Though she speaks approvingly of releasing some repressed human qualities (like sexual desire) that are often considered "dark," Starhawk nevertheless portrays the queer, nonmonogamous characters of her novel *The Fifth Sacred Thing* as uniformly gentle and nurturing in their sexual behavior. Though the novel is based on the real-life Pagan community of San Francisco, that community's full sexual diversity is quietly elided. This omission obscures one of the Pagan community's strengths, its capacity to seek growth through the primal. Many things in nature are frightening: lightning strikes trees, starting raging forest fires; cats catch birds, shaking them hard to break their necks; bacteria, in their task of breaking down dead tissue, also invade the human body to produce deadly illnesses. Yet all these things are a part of God Herself, expressions of the sacred act of creation. How might we contextualize the aggressive or destructive qualities of the natural world?

Theodicy is the study of the problem of evil; classically, it refers to the question of how an omnipotent, omniscient, and benevolent God could allow evil in the world, but it also refers more generally to explorations of the nature of evil and the origins of human suffering. Because Pagan theologies tend to be pantheistic or panentheistic (seeing the divine in the entire world, or the world as completely immersed in the divine), Paganism is often accused of having a weak theodicy—and in some ways, the accusation is fair. Paganism denies that many unpleasant interactions are "evil" at all—the lightning, the cat, and the bacteria all behave in accord with their nature, and the cat and the bacteria act for their own benefit and towards survival as an end. The world is structured so that organisms have competing aims, and this is a thing of beauty and awe as well as pain. Such theologies, however, can lead to confusion about how to critique relations between humans. Is violence just another manifestation of God Herself? Is fighting each other part of human nature, just as it is in our nature to kill bacteria to prevent illness?

Perhaps it is in our nature to fight, but it is equally so to make peace: as a species, we are perpetually engaged in both activities. More specifically, it is the nature of human beings, capable of long-term planning and empathy, to make moral choices. Speaking from the same theological tradition as Starhawk, T. Thorn Coyle asserts that much of what we commonly call "evil" arises from a state of disconnection, a failure to acknowledge our interdependence with the beings that surround us.[48] It is not our capacity to engage in primal and aggressive behaviors that leads us to commit violent acts—for if aggression and hierarchy were the problem, how could BDSM practitioners experience so much joy, pleasure, and connection by playing with them? The assertion that they have "false consciousness" and are mistaken about their experiences is as problematic as telling an adult that she lacks enough awareness to truly give sexual consent. In order to support an individual's right to govern her own body, her report of her experiences must be respected, as Ellison recommends. Although we may seek empathy and connection as a remedy to sin, evil, illness, or pain—however we wish to label those experiences of the world—we do not and cannot read each other's minds; we have no authority to dismiss an individual's own personal narrative and replace it with one of our own.

The attempt to address social and economic inequality by vilifying all hierarchy has not been successful. While I agree with Ellison that deliberately eroticizing peer relationships is a step in the right direction, to eroticize empathic, dynamic power exchange is an equally nurturing and joyful path—one that channels human beings' primal urges and hierarchical tendencies for pleasurable ends rather than repressing or denying them. To understand the world as the body of God Herself and all beings as expressions of hir erotic desire precludes rejecting any part of creation as fundamentally irredeemable. Dominance and submission behaviors exist in cultures around the world and throughout the animal kingdom. From a state of compassionate, erotic connection, it is clear that a society rigidly based on these behaviors—like our own racist, classist, and sexist society—produces unnecessary suffering and deprives many individuals of their right to health and pleasure. Using the lessons of the BDSM community, I suggest that a healthy hierarchy is one in which power flows freely and dynamically, individuals routinely move up and down (and possibly across) existing power structures, and empathy and compassion operate as central values.

Juhan's experience of tapping into society's problems by reading the pain in his clients' bodies points us back to the social importance of physicality. To treat our bodies in a way that produces health, we must remember that we are human animals, guided by our hedonic urges toward both survival and flourishing. Yet we are complex animals, and true health consists of more than the physical. Not only do we have the capacity for moral choice, but we also have a desire for the spiritual health that comes from right relationships, and the need to increase beauty and pleasure both within ourselves and around us. Rather than attempting to leave our animal selves behind, as Western thought sometimes has by prioritizing the mind over the body (as if they could be separated), we have the capacity to embrace our complex urges and, through pleasure, seek to harmonize them: mental, physical, emotional, and spiritual. As self-conscious beings, we are able to redirect and channel our desires; we can discard old strategies of pleasure and health-seeking for ones that better fit our new environment. Once, perhaps, our aggression was needed to keep us safe from predators and to kill the prey that fed us with essential protein and fats. Today, our aggression too frequently results in wasteful wars or the abuse of those around us, including our loved ones. But such feelings can also be channeled: into the grace and control of martial arts, for example, or (with a like-minded partner) enthusiastic play in the bedroom. Aggression and connection need not be at odds—the erotic need not be gentle. When we declare pleasure to be a human birthright and commit ourselves to empathy and compassion in all our relationships, we will move toward greater social health without giving up the primal pleasures that make us human.

Chapter Summary

Beginning with the assertion that pleasure is a human birthright and necessary for health, Chapter 2 explores the ramifications of that belief for social ethics. Chronic stress and touch deprivation have a direct negative effect on human group behavior; clear distinctions between mental/emotional health and bodily health are illusory. Accordingly, as Marvin Ellison points out, social and economic oppression are not mental or emotional abstractions, but are grounded in the devaluing and systematic violation of human bodies. Ellison sees the erotic as a moral force that can drive us to resist such systems of oppression and seek justice. As part of this project, he recommends a collective effort to eroticize social equality.

Against Ellison, I argue that his values of mutuality, intimacy, and diversity might be better served by also eroticizing empathy, the ability to share the perceived emotion of another person. Although relationships without strong power differentials are desirable in many cases, power differentials cannot be entirely erased: individuals can be equal before the law and have equal opportunities, but they will never be equivalent in their strengths and weaknesses. Rather than trying to level the playing field between all potential lovers, I recommend a commitment to dynamic hierarchies where power relations can and do change, both between individuals and between groups.

As a model for this kind of relationship, I engage the sexual ethics of the "safe, sane, and consensual" BDSM movement, in which BDSM practitioners consciously negotiate erotic power exchange. In the process, I reject the argument that anyone who enjoys hierarchical role-playing or intense, even painful physical sensation is mentally ill or caught in a false patriarchal consciousness. Instead, I frame the enjoyment of hierarchy and aggressive sexual play as part of a primal physicality, a wildness that many earth-centered Pagans already embrace in their theology. By conflating all hierarchy and all aggressive behavior with violence, we recapitulate problematic Western tendencies to see the natural, animal world as fallen and evil. An ethics based around the erotic provides a framework for accepting primal human drives and harnessing them in the service of a society based on pleasure, interdependence, personal autonomy, and justice.

CHAPTER 3

Erotic Eclecticism and Divine Deviance

My love, you are a river fed by many streams.
I bless all who have shaped you,
The lovers whose delights still dance patterns on your back,
Those who carved your channels deeper, broader, wider,
Whitewater and backwater lovers,
Swamp lovers, sun-warmed estuary lovers,
Lovers with surface tension,
Lovers like boulders,
Like ice forming and breaking,
Lovers that fill and spill with the tides.
I bless those who have taught you
and those who have pleased you
and those who have hurt you,
All those who have made you who you are.

Starhawk, The Fifth Sacred Thing, pp. 356–357*

Although pleasurable touch is a basic human need, one style of touch does not fit all. As in many parts of life, individuals vary significantly in what they desire. Individual taste is one factor driving the development of numerous bodywork modalities, providing clients with techniques that can be received in many different environments, positions, and states of undress, and with varying levels of movement, pressure, and friction. Strong preferences around touch are perhaps even more pronounced in the sexual realm, where individuals often have not only a favorite set of activities, but particular requirements about the gender and physical appearance of the person who gives or receives the touch. Among progressives, this kind of diversity is usually something to be celebrated: the world's kaleidoscope of languages, art forms, musical genres, religious practices, and ways of being in relationship are seen as beautifully demonstrating the nigh-infinite possibilities for human life. Yet, although acceptable "diversity" has recently come to include some formerly marginalized sexual practices and orientations, many unusual erotic preferences continue to be stigmatized. The embrace of erotic diversity in the United States still falls short of a truly pluralistic erotic ethics.

Envisioning a Pluralistic Erotic Ethics

In her 1984 essay "Thinking Sex," anthropologist Gayle Rubin argues that a pluralistic sexual ethics requires a concept of "benign sexual variation."[1] She writes acerbically:

> A person is not considered immoral, is not sent to prison, and is not expelled from his or her family, for enjoying spicy cuisine. But an individual may go through all this and more for enjoying shoe leather. Ultimately, of what possible social significance is it if a person likes to masturbate over a shoe? It may even be non-consensual, but since we do not ask permission of our shoes to wear them, it hardly seems necessary to obtain dispensation to come on them.[2]

A wide variety of what have traditionally been considered sexual "perversions," Rubin argues, are in fact socially neutral variants in sexual expression, of no more significance than having a preference for a particular sexual position, for candlelight and romantic music, or for redheads. As discussed in the Introduction, this shift in attitudes is in the process of filtering down to the psychotherapeutic community, although it is not yet common in Western society at large.

We have put aside the modernist idea that there might be a single form of government, a single language, or a single philosophical system that will maximize quality of life for the entire human race. Individuals and cultures differ from each other, and it is possible that two competing systems, or the virtues on which such systems are based, can conflict with each other and yet both can be considered good in their contexts. In the Craft, the Charge of the Goddess calls us to have "beauty and strength, power and compassion, honor and humility, mirth and reverence" within us. Yet I do not hear this as a call to the impossible task of demonstrating all these virtues constantly, but rather as a directive to consider them all in making life decisions, and to avoid expressing one virtue while completely excluding its apparent opposite. Within communities—and within the human community—we are called to do the same, collectively expressing paradoxical virtues so that as a species, we can learn more fully what it means to be human.

Contemporary Pagan communities tend to be more accepting of erotic diversity because most Pagans use a pluralistic system of *virtue ethics*, an approach to ethics that in the West can be traced back to ancient Greece. Rather than relying primarily on rule- or consequence-based ethical decision-making, virtue ethics emphasize the cultivation of an individual's moral character within a community context. Virtues such as "compassion" or "courage" are developed as part of a holistic worldview that informs every aspect of an individual's life. These virtues are invoked as guidance when faced with moral decisions, as are role models (such as mythic figures, ancestors, or community elders) who are thought to embody these principles. Since different virtues can suggest different courses of action, contemporary Pagans are challenged to flexibly prioritize their values in a way that contributes to both individual and community well-being. As members of a young religious movement, Pagans struggle to develop the stable communities that can help to ensure ethical accountability and moral development among practitioners. Nevertheless, the Pagan movement embodies a community-based pluralistic ethics that could have much to offer to the wider culture.[3]

I see communities of sexual minorities, particularly those that understand their activities in a spiritual or religious context, as engaged in the process of cultivating particular erotic virtues or values. As with virtue ethics in general, no one virtue will be central to all people at all times in their lives: young people, for instance, are more likely to be in the process of fostering self-reliance and responsibility than elderly people are. Yet the cultivation of a pluralistic erotic ethics offers individuals and communities an entire range of virtuous models of eroticism, one that need not be arranged into an inflexible hierarchy of values. Clearly, the specific practices of alternative erotic communities will not all be relevant or appealing to every individual. Yet the wildness, ecstasy, and culture of enthusiastic consent nurtured by BDSM practitioners; the dualism-defying role fluidity of transgender people; and the abundant intimacy, negotiation skills, and interdependence of polyamorists all offer approaches to eroticism that, modified for various individual and community needs, may offer partial solutions to the oppression, isolation, touch deprivation, and erotophobia that plague many parts of Western society. A pluralistic erotic ethics that accepts these approaches as potentially valid must also necessarily dismantle the strict value hierarchies around "good sex" and "bad sex" that have contributed to the alienation and oppression of women, non-dominant men, racial minorities, and others, not just sexual minorities.

Challenging Rigid Value Hierarchies

In a pluralistic society, it is not enough merely to tolerate diversity; to correct the dualistic thinking that systematically divides Being into rigid pairs of opposites and prioritizes one over the other—male over female, white skin over black, reason over emotion—we must challenge both the idea that the world can be divided so neatly into halves and the necessity of rigid hierarchies of values. Although social theorists have been effectively attacking the assumptions that underlie rigid social hierarchies for decades, these analyses have not been fully extended to erotic and sexual expression. As Rubin writes:

> Progressives who would be ashamed to display cultural chauvinism in other areas routinely exhibit it towards sexual differences. We have learned to cherish different cultures as unique expressions of human inventiveness rather than as the inferior or disgusting habits of savages. We need a similarly anthropological understanding of different sexual cultures.[4]

Many progressives—and I myself have sometimes been among them—struggle to take on this "anthropological understanding" of sexual cultures because sexual variation has been so strongly associated with illness and social dysfunction over the past two centuries, if not more. Michel Foucault writes compellingly about the way Western society attempted to understand and control sexuality with its newfound tools of medical diagnosis and analysis, particularly in the nineteenth century. This *scientia sexualis*, as Foucault calls it, acknowledged the diversity of sexual expression only by pathologizing it. Medical and psychological authorities believed that society was best maintained by privileging monogamous, reproductive sex; only this type of sexual behavior would ensure a stable work force and steady economic growth. Other kinds of sexual behavior threatened to undermine this stable social order, and so authorities created elaborate systems of sexual classification for the purposes of diagnosis, treatment, and "cure."

People who practiced masturbation, spanking, or same-gender eroticism were sorted into categories and labeled as types.[5]

In Shakespearean times, although sodomy was recognized as a sexual practice, it was understood as one (undesirable) behavior among many, not an act that put an individual into a special category. There was no "homosexual" identity until the late nineteenth century,[6] and no gay communities until improved long-distance communication and transportation made it possible for men-loving men and women-loving women to congregate in urban areas. The medical and psychiatric professions, however, have primarily labeled unusual sexual behaviors and practices with the intent of eradicating them. (Rubin names empirical sex researcher Alfred Kinsey as an exception, as his research beginning in the 1940s catalogued and described the sexual behaviors of Americans with an unjudgmental eye that did not conflate an individual's personality with their enjoyment of particular sexual acts.[7]) Until recently, mainline religious communities largely followed the lead of medical professionals, leading to the mass exodus of LGBT and queer people from mainline churches over the past few decades (and the growth of more accepting religious communities such as contemporary Paganism).[8]

Tolerating Queerness

As discussed in the Introduction, theologians, activists, and theorists have recently begun to use the term "queer" to include not just LGBT people, but also those with unusual sexual practices and heterosexuals who are ideologically allied with the queer community. In *Tendencies*, theorist Eve Sedgwick writes poetically:

> Queer is a continuing moment, movement, motive—recurrent, eddying, *troublant*. [. . .] The immemorial current that *queer* represents is antiseparatist as it is antiassimilationist. Keenly, it is relational, and strange.[9]

Definitions like Sedgwick's suggest that to be queer is to be inherently subversive, boundary-breaking, and troublesome. That which is queer cannot be assimilated into a group, nor can it separate from the group entirely; it challenges social norms, yet endlessly dialogues with them.

At the April 2007 "Queering the Church" conference at Boston University, theologian Marcella Althaus-Reid's lecture[10] characterized queer theology as essentially committed to flux and newness. Like process theology, queer theology rejects the idea that there can be any eternal, disembodied truth apart from change: all of Being is embodied and changing, and therefore diversity is endless. In response, theologian Carter Heyward described the image of the biblical God's infinitely spacious house as a queer image. In the gospel of John, Jesus tells his disciples, "In my Father's house there are many dwelling places. If it were not so, would I have told you that I go to prepare a place for you?" (John 14:2 New Revised Standard Version). For Heyward, this verse describes God's house as having an infinite capacity to expand and encompass endless possibilities. This indiscriminate acceptance (or *grace*) is queer, she argues, because of the way it ignores social norms of decency and propriety, just as Jesus did when he associated with prostitutes, tax collectors, and lepers.

BDSM fits this expansive definition of "queer" as process-oriented, challenging, and subversive, as does contemporary Paganism. Yet the advocacy of BDSM and Pagan spokespeople for greater acceptance threatens to render these alternative paths "normal"—an assimilation into society that will rob them of their most valuable contribution, the insights provided by a marginalized position. For that reason, my advocacy for these minorities communities and for a pluralistic erotic ethics is somewhat nuanced. I would like to see a world where same-sex and multiple-partner relationships are indeed normal, and where having unusual sexual tastes does not put people at risk for loss of jobs or child custody. I would like to see transpeople able to easily access psychological and medical services and to make decisions about their gender expression without being disowned by their families or rejected by friends. I would like members of all these groups to be invited to the table when community groups and organizations make policies about sexual ethics. And I wholeheartedly advocate the spread of ecologically-aware, nature-oriented spiritual practices that might help to slow or reverse our abuse of the natural environment.

Ultimately, however, my concern for the more radical minority communities (such as the BDSM "edge players" described by Easton and Hardy and those Pagans who seek altered states through extreme physical means) is about *preserving their civil liberties and ensuring a minimum level of social tolerance.* The assimilation of the queer is not only impossible, but undesirable: a society cannot fully include the outsider without losing the outsider's valuable perspective. Yet when it comes to consensual practices between adults, we could go much further in cultivating respect for the necessary outsider, the cultural innovator or risk-taking spiritual explorer who violates social norms. I do not believe that BDSM, gender-bending behavior, or ethical nonmonogamy will be positive, healthy erotic practices for every individual. But I do believe that we must end the demonization of sexual minorities, and that all people can learn from listening to the insights of these marginalized groups.

Anti-Porn Feminism and Lesbian BDSM

Although many of us have come to understand masturbation and same-sex desire as part of the normal range of human sexuality, enjoyment of group sex, BDSM, fetishism (like the aforementioned sexual enjoyment of shoes), and pornography are still often condemned as damaging to society. Much of the rhetoric condemning BDSM and sexually explicit art today can be traced back to the anti-porn segment of the feminist movement of the 1980s. Without a doubt, some pornography objectifies women, men, and transpeople, and the industry is in desperate need of regulation. The anti-pornography crusaders of the 1970s and 1980s, however, sought their vision of women's liberation at the expense of other marginalized groups. Radical sex writer, therapist, and Pagan Pat Califia writes of her ejection from the lesbian feminist movement in the late 1970s because of her practice of BDSM. (During this period, Patrick Califia identified as a lesbian rather than as a transgender man. Since lesbian identification is such an important theme in this period of his work, I refer to him in historical context using feminine pronouns.) Writing as a lesbian, Califia chronicled her coming-out as a BDSM practitioner in the lesbian community of San Francisco. Unfortunately, she and her allies experienced rejection from much of the rest of the lesbian feminist community, since many lesbian feminists saw BDSM practice as inherently patriarchal. For a committed feminist like Califia, this experience was doubly painful: lesbian feminists were being actively rejected by mainstream culture, so with the additional rejection of the feminist community, kinky lesbian feminists became a hated minority within a hated minority.[11]

As members of the lesbian sadomasochist (S/M)[12] movement, Califia and Gayle Rubin were denied admission to the first national anti-pornography conference put on by Women Against Violence in Pornography and the Media (WAVPM). Their lesbian S/M group, Samois, was accused by WAVPM of glamorizing violence against women.[13] Although skeptical of the idea that pornography is inherently violent, Califia and others from the group arranged to attend WAVPM's educational slideshow on the topic after WAVPM refused to address Samois directly. Califia reports:

> When I saw the slideshow, I was frightened by the sexual ignorance of the presenters and their willingness to exploit the audience's prejudices to win converts.
>
> The images were presented in a manipulative way. Bondage photos were followed by police photos of battered women. S/M porn was repeatedly taken out of context and labeled as "violent,"

> even though it looked like consensual S/M to me. Record album covers and fashion advertisements were lumped in with the pornography. Nothing was dated and few sources were cited, so the audience had no idea how representative the material was of porn you could actually go see and buy. The presenters kept assuring us that much, much worse material was available. Their definition of porn was circular and sloppy. They defined any sexist or violent image as pornography, then turned around and used that assumption to "prove" that all pornography was violent and sexist.
>
> Lesbian sexuality was not discussed. Some vague distinction was made between "erotica" and "porn," but no examples of "erotica" were shown. This made me especially uncomfortable since many heterosexuals were present, and one of the favorite images of "violent" pornography was soft-core, glossy images of women kissing or going down on each other. [. . .] There was no sense of context or scale. The lesbian porn was presented as being just as violent as a woman getting stabbed. Wearing high heels or being tied up was described with as much horror as getting raped. Corsets were condemned with as much vehemence as wife-beating. Anal sex was apparently a violent practice. [. . .] Anybody who questioned WAVPM's definition of porn or violence was accused of having bad consciousness about violence against women.[14]

Califia's essay was first published in 1981. In the decades since, anti-porn feminists have lost much of their status among progressives and in the feminist movement as a whole, largely due to the lack of hard evidence that consuming pornography influences viewers to assault or abuse women.[15] The rise of an industry to produce pornographic films and images by and for women of various sexual orientations has also undermined the claim that enjoyment of pornography is a sickness resulting from patriarchy. Although much pornography is tasteless and objectifying to the individuals in it, those looking for erotic entertainment today will be able to find materials that show people of all genders enjoying their sexual power, as well as interviews with porn stars speaking about the rewards of their work.[16] While there are still clearly exploitative aspects to the pornographic media industry—aspects that affect all genders, though women and transpeople still bear the greater brunt of the impact of institutionalized sexism—the claim that pornography is *inherently* damaging to women no longer seems convincing to most third-wave feminists, for whom sexual liberation is a primary feminist goal.

Sexual Panics and Social Control

The attack on pornographers and practitioners of BDSM by a certain contingent of the feminist movement is an example of what Rubin calls a "sexual panic," a term that describes how anxieties around deviant sexuality have been used both historically and in the present day to pass legislation enforcing dominant social, economic, and gender norms. Drawing on Jeffrey Weeks's research on moral panics, she writes:

> Sexual activities often function as signifiers for personal and social apprehensions to which they have no intrinsic connection. During a moral panic, such fears attach to some unfortunate sexual activity or population. The media become ablaze with indignation, the public behaves like a rabid mob, the police are activated, and the state enacts new laws and regulations. When the furor has passed, some innocent erotic group has been decimated, and the state has extended its power into new areas of erotic behavior. [. . .] The target population suffers most, but everyone is affected by the social and legal changes.[17]

Writing in 1984, Rubin predicted that the political right would use AIDS to encourage homophobia and combat the gay rights movement, an expectation that history has borne out.[18] The attacks of anti-pornography feminists on sadomasochists also tended to specially implicate gay men, as the most visible sadomasochists in the early 1970s were the gay leather groups centered in San Francisco.[19] The visibility of gay sadomasochists—and the existence of lesbian ones—seemed not to figure in to the anti-porn feminist claim that sadomasochism is associated with men raping women, however. (According to Rubin, the association of BDSM and other unusual sexual practices with rape is unsupported by sociological data—so-called "sexual deviants" do not disproportionately commit sex crimes.[20]) Rubin also observes that the rhetoric used by feminists to pass legislation condemning pornography and BDSM has played into the hands of sexual reactionaries. Pope John Paul II, for example, employed sexual objectification arguments used by anti-pornography feminists in order to condemn divorce, abortion, and birth control.[21] It is all too easy for one oppressed or underprivileged group to gain power and recognition for itself by directing social anxieties at yet more oppressed and persecuted people—often without realizing that their own tactics may be turned against them by those in power.

But to return to the issue of the pathologizing of minority sexualities in our culture, it is not hard to see why progressives concerned with the liberation of women might initially look askance at BDSM practices. Sexual abuse and assault are disturbingly common crimes in our culture, and they are suffered by adults and children of all genders. Without ignoring the fact that men are also sometimes the victims of these crimes, however, women are molested at higher rates, with some studies suggesting that 20–25 percent of all American college women experience attempted or successful rape during their college years.[22] (Those who argue that BDSM should be illegal because extreme acts are physically or emotionally "riskier" should take note that in terms of sheer numbers, heterosexual intercourse is the sex act most often used abusively in American society—and yet few people would argue that heterosexual intercourse should be banned as a result.) Particularly among those who are unfamiliar with the "safe, sane, and consensual" model of BDSM, the idea that a woman might *want* to be tied up and beaten can seem like a mental illness caused by the harsh realities of patriarchy: since men rape women every day, perhaps some women, poor souls, have begun to mistakenly believe they enjoy and desire this subjugation.

Granting Moral Complexity to "Deviant" Sex

As Carter Heyward suggests, it is possible that if our society did not have strong hierarchical tendencies, and if human beings did not regularly abuse and violate each other, there would be no such thing as BDSM sexuality. But I would argue, as I did in Chapter 2, that BDSM sexuality is less a symptom of society's illness than an attempt to effect a cure. In the kink community, people of all genders play top and bottom roles and enjoy giving and receiving intense sensation (that is to say, women, men, and transpeople can be found in both top and bottom roles, although individuals' willingness to switch roles varies by community). In BDSM play, the participants' primal selves are invited out to play in an environment of carefully negotiated trust. What would be abuse if the recipient were not consenting becomes an exhilarating exploration of one's own emotional and physical limits. As a friend and rape survivor told me recently, it is incredibly sexy and exciting when her partner trails the tip of a knife along her skin not because she believes she might be cut, but because it dramatizes her vulnerability and the knowledge that she is perfectly safe and deeply loved.

Rubin argues that in Western society, only certain forms of sexuality are considered acceptable and are therefore accorded "moral complexity": "For instance, heterosexual encounters may be sublime or disgusting, free or forced, healing or destructive, romantic or mercenary [. . .] In contrast, all sex acts on the bad side of the line are considered utterly repulsive and devoid of all emotional nuance."[23] Because BDSM practice is frequently pathologized or misunderstood as violence, practitioners often keep their predilections even more private than most people keep the details of their sex lives. While it is usually not considered polite to discuss one's bedroom activities while at work or with one's parents, BDSM practitioners refrain not just as a matter of social etiquette, but because being outed can result in losing a job or worse, losing custody of their children.[24] The assumption of pathology around BDSM obscures the fact that most BDSM practitioners lead lives that are normal in most respects. BDSM practitioners are doctors and artists, computer programmers and construction workers; they have partners and children; they attend religious services, have picnics, cry at sad movies, and otherwise generally act like healthy, compassionate human beings. As with other communities, of course, a minority of BDSM practitioners are emotionally or physically violent in their relationships, neglect or abuse their children, and experience mental illness. Researchers and therapists have repeatedly argued, however, that violent behaviors are no more common among BDSM practitioners than in society as a whole.[25] Nor are these behaviors curbed in sexually conservative environments. Sex advice columnist Dan Savage, by way of rebutting the claim of the Christian right that openly gay men's sexual orientation causes them to prey upon and abuse children, has made a project of collecting news stories about child abuse, rape, and sex crimes in church communities and by conservative Christian spokespeople.[26] Savage's hobby is sadly easy to pursue, particularly with regard to the Catholic Church, whose systematic efforts to cover up incidents of child abuse by priests made headlines in the first decade of the 2000s.[27] This anecdotal evidence of abuse, however, does not lead me to believe that sexually repressive environments necessarily encourage acts of sexual violence or exploitation either—only that erotic conservatism does not prevent it.

Since the time of Rubin's writing in 1984, the taboo against homosexuality has weakened to the point where healthy same-sex relationships can be routinely seen on television and in movies; in some cases, these include positive representations of same-sex eroticism. BDSM relationships are far from reaching that point of acceptability, although the recent success of the pulp erotic novel *Fifty Shades of Grey* has made them more visible. Perhaps because of the strength of the social taboo against kink, BDSM practitioners have come to identify their sexuality as an orientation and themselves as a distinctive sexual subculture in need of advocacy, as gays and lesbians once did. Gay pride parades now sometimes feature men, women, and transpeople wearing black leather and waving black- and blue-striped flags, signaling "leather pride" in the hope of creating enough visibility and acceptance that they might be assured of their civil liberties.

Certainly there are BDSM practitioners in the world who have kinky sex for the sheer pleasure of it, just as there are many heterosexual couples who enjoy their sex lives without understanding sex as a religious practice. As discussed in the Introduction, however, some BDSM advocates make a case that all sex is a potential path to divine experience—and kinky sex may actually be more so, due to its including techniques used to induce ecstatic states in mystical traditions around the world. In *Radical Ecstasy*, Janet Hardy and Dossie Easton present themselves as long-time friends and lovers, now in late middle age, who have found BDSM to be an essential part of both a decades-long sexual relationship and a fulfilling spiritual practice. Through physical ordeals and sexual roleplaying, Hardy and Easton have sought not only joy and pleasure, but altered states of consciousness similar to those attained in religious practices such as yoga or tantra. They write:

> So what is this bliss, this ecstasy, that we seek—through religion, through meditation, through sex, through SM? Definitions are always too small to fit the amazing reality we experience in ecstatic states. This ecstasy, as we and many others who write about it experience it, adds up to experiential proof that the universe is bigger than us—*lots* bigger. And if some people want to call the universe God or Goddess, there's nothing wrong with that, as long as we understand that God or Goddess is a name for something so immense that we cannot in any way begin to describe it[.][28]

Easton and Hardy's views are not uncommon among Pagans. In her book *Rites of Pleasure: Sexuality in Wicca and NeoPaganism*, Jennifer Hunter interviews twenty-two members of the contemporary Pagan movement—some well-known authors and leaders, some simply participants—and combines their viewpoints with her own several decades of experience in that community. Chapter 6, "BDSM and Dark Eros," explores BDSM as a shamanic spiritual practice and articulates a similar point of view to *Radical Ecstasy* (Easton, in fact, is one of the interviewees). The majority of interviewees discuss BDSM as a path of self-discovery and healing, although one respondent cautions that (as with intoxicants) the highs gained through BDSM can be addictive; another warns against the possibility of hiding emotional abuse beneath "safe, sane, and consensual" BDSM rhetoric.[29] Significantly, however, Hunter's survey of BDSM in the Pagan community reveals a range of emotional possibilities. Certainly one can commit acts of violation and violence during BDSM play, just as violation and violence are possible during heterosexual intercourse. But what society as a whole does not accept about sexual deviants is that their capacity for emotional and moral complexity is just as great as those who only have "normal," socially-condoned sex. The prejudice of mainstream society against alternative sexualities is revealed in the fact that those outside the mainstream are *assumed* to lack moral integrity. Rubin counters,

> This kind of sexual morality has more in common with ideologies of racism than with true ethics. It grants virtue to the dominant groups, and relegates vice to the underprivileged. A democratic morality should judge sex acts by the way partners treat one another, the level of mutual consideration, the presence or absence of coercion, and the quantity and quality of the pleasures they provide.[30]

Perhaps it is true, as those who have crusaded against pornography and BDSM have claimed, that what is done in the bedroom affects the whole of society—the personal is political, to echo the feminist slogan of the 1970s. But as I suggested in Chapter 2, it is the *emotional resonance* of what happens in the bedroom that affects the whole of society, not the outward forms. The details of a couple's sex life are far less important than the honesty, respect, and empathy that they demonstrate toward one another. If an erotic act is experienced as loving by both parties, then it is love that will be felt by the couple's community, not whatever emotional content the community might project onto the act. A pluralistic erotic ethics understands that sexual pathology is located in the participants' *experience* of a sexual act rather than being inherent in the act itself. Heterosexual sex can be holy, or it can be rape. Sexually "deviant" acts should be accorded the same complexity.

Destabilizing Moral Dualisms

Previously, I mentioned the liberation theology notion of a "preferential option for the poor" and suggested a parallel for sexual deviants—a preferential option for the queer, so to speak. Those who find themselves on the margins of society due to their sexual preferences have special contributions to make to the human conversation about sexuality, both due to their position outside of structures of privilege, and because their tastes may give them opportunities for unusual, challenging, and perhaps growth-inducing experiences. I have already spoken to some extent about the special gifts of the BDSM community: BDSM practice provides a negotiated, considered method for experiencing the primal, aggressive desires that are the inheritance of our animal bodies. Further, dominance and submission play allows BDSM practitioners to subvert standard social power differentials or to explore their ramifications and consequences within strict limits, all for the purposes of discovery and pleasure. Easton and Hardy, however, emphasize BDSM as a route to mystical experience that they believe can encourage compassion and empathy.

It has become commonplace among progressive thinkers to speak disparagingly of strict Western philosophical dualisms: because we have seen the way that dividing the world into strict binaries (male/female, reason/emotion, civilized/savage) can contribute to exploitative social hierarchies and an inability to recognize complexity, we reject black-and-white thinking and instead wish to see the world in more complex terms. To oppose this kind of dualistic thinking intellectually, however, does not always prevent us from falling into it when our emotions and sense of justice are aroused. American national policy on issues such as environment, health care, and defense are matters of life and death for ourselves and our children, and it is all too easy to slip into problematically polarized thinking and to see social interactions as a war between opposing sides (conservative vs. liberal, red vs. blue, etc.). Given the vision of underlying spiritual unity in Starhawk's *The Spiral Dance*, I was disappointed to see her implicitly affirm polarized political divisions on the occasion of the book's thirtieth anniversary:

> I see two roads for the future—and that's part of the theme and imagery of this year's and previous Spiral Dances. On one road, we continue to pump fossil fuels into the atmosphere and pump the poisons of fear, racism, hate, and war-mongering into the psychic atmosphere. By 2039, we'll face a world of drought, famine, endemic war, potentially a loss of our civil liberties, hundreds of millions of deaths, oceans rising . . .
>
> Then there's the other road, the good road, the road of life . . . where we make the tremendous shifts we need to make, where we recognize the sacred in every human being and in the interconnected web of all life, where—as our litany says—"we draw our power from the wind and sun."[31]

While I share elements of the utopian vision that Starhawk calls "the good road," I find the implication that supporting solar and wind power goes ideologically hand-in-hand with peace and tolerance to be highly questionable. The social, political, and environmental issues to which she refers are complexly related, and it is possible for compassionate and thinking people to hold a range of opinions about their proper solutions. To characterize society as being on the brink of choosing either totalitarianism and environmental collapse or an egalitarian, ecofriendly utopia lends itself to us-versus-them thinking and the demonization of political moderates and conservatives. Such polarizations make the consensus-building necessary to set policy and move forward on socially beneficial projects nearly impossible, even on the more manageable level of local politics.

Pagan writer and theologian John Michael Greer observes that problematic dualisms often underlie allegedly nondualistic Pagan and earth-centered theologies. In examining Pagan origin myths (such as the historical narrative of a pacifistic, Goddess-worshipping prehistory that Riane Eisler describes[32]), Greer finds the same kind of moral dualism that underlies the strictest monotheistic theologies:

> There are two sides, and only two; one is right, and the other is wrong. There is no middle ground, no moral ambiguity, only good and evil in stark contrast.
>
> [. . . T]he myth is *agonistic*—that is to say, it's a myth of war. [. . .] Peace and harmony are restored only when one side no longer exists.[33]

In the work of Eisler and other feminist writers, there is a perceived culture war between the Neolithic pacifistic worshippers of the Goddess (and those contemporary people who honor the feminine principle) and the patriarchal, hierarchical, warlike powers that conquered Old Europe long ago (and whose political heirs hold the reins of power today). Greer observes that for such writers, there is no proper or healthy place for those whose values differ from the "good" side; the opposing side must be (peacefully?) wiped out through cultural change. He continues, speaking of the writing of earth-centered Christian theologian Matthew Fox, whose writings are admired by many Pagans:

> [Fox's] argument—as presented at length in *The Coming of the Cosmic Christ* and other books—is that there are two and only two kinds of religion: dualist and nondualist, which are utterly opposed to one another. Dualism is absolutely evil, while nondualism is absolutely good. At one point he spends the better part of two pages running through a long list of polar opposites, defining one ('nondualist') pole as good and the other ('dualist') as bad[. . .]. Pick up books by Starhawk, Riane Eisler or any of several dozen others, and you'll find ringing critiques of dualist thinking phrased in highly dualistic terms.[34]

I can only speculate as to why those supposedly committed to honoring all things as sacred nevertheless fall into dualistic intellectual traps. Dualisms may persist problematically in politics and religion because other kinds of rigid binary thinking have been so helpful to our survival: practical binaries such as edible/poisonous, safe/dangerous, or friend/stranger. In contemporary life, however, an overabundance of dualistic thinking can preclude awareness of the diversity of options available to us politically, philosophically, and socially.

Interestingly, Hardy and Easton offer BDSM as one practice that can loosen the hold of dualism, not through theory and argument, but through physical and emotional experience. They write:

> Transcendent [sexual or SM] play is a way that we dissolve the boundaries of space and time, and the walls that seem to keep us apart from the people we care about. Where do those boundaries, those barriers, come from? Sometimes, we think, they come from patterns we learn when we decide that a thing must be either one way or another. If the divine is in everything, the divine is both one way and another: as simple, and infinitely complex, as that.
>
> When you start out by cultivating your tolerance for ambiguity and paradox, you are loosening the strings on your mind, giving yourself permission to feel without judging, to trust your sensations and emotions rather than your busy brain. You are practicing believing in what you feel at the moment, not reflecting the past or fretting about the future. You are practicing for ecstasy.[35]

Easton and Hardy characterize BDSM play as something that allows them to see the world more accurately. Positive ecstatic, boundary-blurring experiences raise one's tolerance for the ungraspable and uncategorizable, and so potentially permit us to re-examine our assumptions, rather than too-quickly fitting new information into an existing worldview. Further, in loosening the hold of dualistic thinking on our minds, we become better able to willingly drop our barriers and experience ecstasy. BDSM, like meditation or tantric sexual techniques, is one method of conditioning oneself to experience altered states of consciousness. With each ecstatic experience, one becomes a little more open to the "paradox and ambiguity" that characterize human experience—and in turn, with each additional degree of tolerance for that paradox and ambiguity, ecstasy becomes a little easier to achieve. Like an experienced practitioner of Buddhist meditation who may slip into a state of blissful oneness within minutes of beginning her morning sit, experienced BDSM players can condition themselves to experience states of consciousness that help them to lead happier, more grounded, more compassionate lives.

Easton and Hardy's belief that BDSM can create opportunities to modify formerly rigid worldviews is borne out by research into the role of pain in religion. Elaine Scarry, for example, portrays pain as something that causes a human being to revert to a state before language and forces the recipient to modify existing mental constructions of the self and the world. Such experiences, she argues, can drive valuable religious and cultural innovations.[36] Similarly, Ariel Glucklich uses neuropsychological analysis of pain to argue that experiences of pain can alter or destroy a person's former goal orientation or sense of self—in other words, a kind of death-rebirth experience that in a religious context can be used to reorient the participant toward spiritual goals.[37] Hardy and Easton's spiritual goals include the creation of an egalitarian, pluralistic society. In their own lives and in their communities, they use the practice of harnessing intense physical sensation to loosen dualistic cultural constructions that they believe underlie systems of oppression and violence.

Shamanism and the Collapse of Gender Binaries

The dualistic thinking that orders power according to perceived gender is under attack in Western society, but gendered assumptions still hinder individuals' careers, lifestyle choices, modes of dress, and more. Pagan shaman Raven Kaldera sees transgender people as inherently threatening to these persistent gender binaries. In his book *Hermaphrodeities*, Kaldera retells the Greek myth of the hermaphroditic deity Agdistis. Feared by the gods because Agdistis had access to both male and female magics, the gods first demand that Agidstis choose a gender, then castrate hir by force. Kaldera points out that this myth is replayed in the systematic mutilation of babies born with intersex conditions, whose genitals are altered in order to give them "normal" gender identification.[38] Unfortunately, this procedure is done without knowing how the child will ultimately express hir gender. Anecdotal data suggests that in many cases, nature trumps nurture. One famous case is that of David Reimer (originally Bruce/Brenda), who was raised as a girl after a botched circumcision ended in his penis being removed. The result was a persistent gender dysphoria that ended when, at age fourteen, he learned about his history, stopped taking estrogen supplements, and assumed a male name.[39] Although David's intersex condition was created by doctors, the same dilemma exists for babies born with two sets of genitalia. Even more complex are the subtler intersex conditions, such as having both ovaries and testes, which may not be detected until adulthood, if at all. These conditions are not as rare as most would think: studies estimate that up to 4 percent of the general population has an intersex condition of some kind.[40] Not all transgender people have an obvious biological reason for gender dysphoria, however, and solutions to the discomfort many experience include social, psychological, and biological changes. Individuals may find relief in alternative gender expression, or through altering their bodies with hormone treatments or surgery. However, Kaldera notes that, as in the myth of Agdistis, transgender people are frequently targeted for violence, some of which aims to "correct" their gender transgression.[41] (Based on a true story, the 1999 film *Boys Don't Cry* portrays the transgender Brandon being gang-raped by his former friends after they discover his female genitalia—the intention being to teach him that he is a woman.) Transgender people often live in fear of violence, as the revelation of their unconventional gender may trigger intense fear or rage in those around them.

Transgender people cope with their outsider status in a variety of ways. In *Gender Outlaw*, transgender writer and advocate Kate Bornstein speaks of two destructive threads of thought within the transgender community: the tendency of some to consider themselves "the chosen people," and the contrasting urge to see themselves as abject victims.[42] Caught in the liminal space between social categories of male and female, transpeople suffer the brunt of gender anxieties in our society and are frequently beaten and even killed for their difference.[43] The deaths of transpeople may serve a scapegoating function for our society, which is intensely anxious about the blurring of traditional gender roles and the instability of the institution of the nuclear family. The term "scapegoating" comes from a biblical practice in which the Israelite community symbolically piled its sins on a sacrificial goat and then drove it away to die. When transpeople die at the hands of transphobic assailants, like the sacrificial goats, they are innocents being punished for our society's collective dysfunction about gender roles. Today in my home city, it is safe for me to go out clubbing in a gender-bending outfit—an act that would have been physically dangerous a mere forty years ago. My safety is due in part to the awareness raised by transpeople, individuals who risked their lives to express themselves authentically in public.

Bornstein sees the position of transgender people as both a vulnerability and a strength. Unable to fit into standard social categories of gender, transpeople are also unbound from the restrictions of that system, with its rules about how proper men and women behave in the workplace, with their families, or in their bedrooms. Yet transpeople do not choose gender dysphoria, so their status as "gender outlaws" is at least partially forced upon them. Like Kaldera, Bornstein sees transgender people as being unwillingly pushed toward a shamanic role in our society. "Shaman" is an anthropological term for a person who contacts the spirit world on behalf of his community. The process of becoming a shaman frequently involves a spiritually and physically difficult transformation in which the initiate symbolically dies, then returns to society to be of service as a healer or oracle. Socially marginalized and frequently persecuted, transgender people often sacrifice their ill-fitting gender identities in a painful ritual that can cost them friends, family, and career and may involve a dangerous surgical transformation. Although some transpeople fully integrate into mainstream society after transition, the transition experience often undermines or destroys their confidence in dual-gender social systems. Such an altered perspective can be employed in service of the wider community—but for most transpeople, acquiring such a perspective is not a choice, but rather a result of mainstream society's rejection.

Bornstein observes that the vast majority of transgender people do not feel that they chose their path. Some transsexuals—those who wish to or have transitioned from male to female, or female to male—believe that their condition has a biological basis and so is best addressed through biological means (hormone treatments and/or surgery). Current research also suggests that there are biological determinants for sexual orientation.[44] Although it is possible, to various degrees, to choose a leather, polyamorous, or otherwise sexually deviant "lifestyle" by connecting with communities of like-minded people, many of those within those communities feel that there is a biological component to their sexual desires, even if these desires have been shaped by the culture in which they live. Although chromosomes and hormones may not be sufficient to create a queer perspective, it is clear that for some sexual minorities, sexual and gender difference begin with biology.

Acknowledging the importance of biology in sexual and gender difference, however, need not lead to rigid or essentialist thinking about the subject. The feminist Pagan theology of Constance Wise, for instance, provides a way to understand gender through a framework of process thought. Like all of Being, she suggests, gender roles (and potentially gender identities) arise from complex physical, mental, and spiritual processes and may change over time:

> Each individual experiences life as a gendered person. Individuals take in the realities of gender partially through the norms and messages of their social context, partially through their own life experiences, and partially through the creativity of their own personal interpretation of gender. Each person communicates these experiences to others so that every woman, man, and child receives, interprets, and contributes to the open-ended reality of gender.[45]

Although Wise does not directly address transgender issues, her notion that gender is an "open-ended reality" and her intention to break through "essentialistic gender expectations"[46] resonate with the concerns of the transgender community. Process thought is inherently anti-essentialist: because of its emphasis on contingency and change, it is incompatible with the idea that concepts such as "deity" or "womanhood" have inherent, necessary, unchanging characteristics (in other words, they have no eternal essence). In process thought, gender is one of many concepts that is constantly being constructed, changing based on social, biological, and political conditions. What we understand as "womanhood," "manhood," and "gender transgression" have historical precursors and connect to physical realities, but they nevertheless evolve over time.

Regardless of the basis for one's gender identity, sexual orientation, or erotic desires, however, it is clear that individuals' well-being and happiness depend upon the ability to live in a way that feels authentic. Since at least the 1970s, Pagans have been actively retheorizing gender by rewriting conventional gender roles in an egalitarian framework and, more recently, sacralizing specifically transgressive or queer approaches to gender. Today, the subject of gender and its role in Pagan practice remains a topic of passionate debate.[47]

Polyamory and Erotic Abundance

After the sexual revolution of the 1960s and 1970s, the notion of ethical nonmonogamy may be far less off-putting to progressives than the more unusual or extreme erotic practices described elsewhere in this book. Those of us who choose monogamous relationships can still often understand the appeal of being able to have multiple lovers without breaking vows or deceiving a loved one. Yet the tendency to pathologize remains, as those polyamorous families who have gone to court to retain or regain custody of their children can attest.[48] Believing that the unorthodox sexual relationships of the parents are harmful to the children's well-being, grandparents and other blood relations have sometimes successfully pursued legal action to remove children from their families. Even those with sympathetic and curious attitudes toward polyamorists and practitioners of group sex have sometimes seen the behavior as a result of imbalance or illness. Researcher Arno Karlen's 1988 study *Threesomes* gathers twenty years of interviews with and observations of those who engage in group sex, some of whom live in ethically nonmonogamous family arrangements. Though relatively unjudgmental in tone, Karlen's work suggests that those who engage in threesomes are compensating for unsatisfying family lives in childhood. The threesome, he says, provides a unique experience of group intimacy that addresses unmet childhood needs. Although Karlen sees the experience of group sex as psychologically revealing and potentially deeply meaningful, his argument implies that if the childhood families had been healthy to begin with, participants would find multiple sexual relationships unnecessary.[49]

Unapologetic supporters of ethical nonmonogamy also share the idea that nonmonogamy and the desire for intimate family ties are linked, however. For example, Pagan writer Raven Kaldera does not portray the desire for multiple sexual relationships as a sign of ill health, but rather as a natural desire for more joy. His book *Pagan Polyamory* presents ethical nonmonogamy as an attempt to reinvent tribal extended family structures, in which individuals develop complex networks of lovers, ex-lovers, and lovers-of-lovers. To illustrate, he quotes a letter from Pagan leaders Oberon and Morning Glory Zell-Ravenheart, who write:

> The freedom of having more than one devoted, bonded relationship is a joy that is almost impossible to describe to someone who has not experienced it. There is an inspiration to it, and amazing security. To us it is a human triumph of communication skills, moxie, romantic inspiration, and flexibility.[50]

In a statement that resonates with the pluralistic virtue ethics used by many Pagan communities, Kaldera also draws parallels between polyamory and polytheistic theology. If one is able to honor more than one deity or aspect of the divine in one's religious life and can benefit from the differing virtues of those deities, then to be devoted to and benefit from having more than one lover is similarly both possible and desirable.[51]

Polyamorists have special contributions to make to our culture's ongoing conversations about sexuality. As with BDSM, which traces some of its roots to traditional tribal practices of body modification and ordeal rituals, polyamorists' ideas have historical precursors in unlikely places. For instance, contemporary polyamorists have much in common with John Humphrey Noyes, the founder of the devoutly Christian nineteenth-century Oneida community. The Oneida community prioritized their relationships with God through a number of unusual lifestyle choices, including the practices of communal property and plural marriage. Noyes writes:

> The law of [monogamous] marriage "worketh wrath." It provokes to secret adultery, actual or of the heart. It ties together unmatched natures. It sunders matched natures. It gives to sexual appetite only a scanty and monotonous allowance, and so produces the natural vices of poverty, contraction of taste, and stinginess or jealousy. It makes no provision for sexual appetite at the very time when that appetite is strongest . . . This discrepancy between the marriage system and nature is one of the principle sources [of the sexual tensions and disorders of men and women] . . . The restoration

> of true relations between the sexes is a matter second only in importance to the reconciliation of man to God.[52]

Noyes believed that ethical nonmonogamy would restore right and natural relationships between the sexes, result in greater sexual and emotional satisfaction, and permit individuals to prioritize their relationships with God rather than unhealthily fixating on a single romantic partner. While most contemporary polyamorists would probably disagree with the reasoning of the last point—the heady passion of NRE, or "New Relationship Energy," is one of the dangerous delights of a polyamorous lifestyle—the freedom and joy of which Noyes speaks are important values for polyamorous communities.

Unlike swinger communities, in which participants tend to keep outside sexual relationships separate from a primary love relationship, polyamorists seek to stuff their lives full with pleasurable activities with loved ones, not all of which are sexual. I have been fortunate to have had a long personal relationship with the Boston poly community, which is one of the largest and most mature of its kind in the United States. There, one can meet polyamorous families whose children have grown up to practice polyamory themselves; child-friendly gatherings are common, and children are entertained and cared for not just by their parents, but by extensive "chosen family" networks of friends, lovers, former lovers, and other loved ones "too complicated to explain." To a polyamorist, if one happy relationship is good, three are certainly better; although some of these relationships are more casual dating relationships, it is not unusual to meet individuals who have had two committed partners for a decade or more.

Unsurprisingly, polyamorous communities suffer from overstuffed social calendars and a tendency to overcommit. Jokes about scheduling snafus are common, and setting a date a month in advance is a normal practice. Further, the communication and negotiation necessary to maintain such complex webs of relationships take a great deal of skill, patience, and work. Although as marriage counselors and therapists know, healthy monogamous marriages also require a great deal of explicit expectation-setting, boundary management, and negotiation, polyamorous relationships require this level of commitment and care among many relationships at once. Those who are prone to strong jealousy do not generally stay long within the polyamorous community, but some amount of jealousy is considered normal and must be managed. Successfully polyamorous individuals are skilled at stating their own needs and wants, but are also accustomed to taking responsibility for their own self-care. Those who have not experienced polyamory may easily observe the warm network of connections found in such relationships, but fail to notice the emotional independence and assertiveness that is required of each individual involved.

The communication and negotiation skills that are considered normal in the polyamorous community are often not developed in mainstream monogamous relationships until a crisis arises. Only then, with the advice of good friends, self-help books, or a counselor do couples learn to fight fairly and take responsibility for their own needs. In polyamorous relationships, however, these skills are necessary to get most relationships past the first month or two of life. (Indeed, among newcomers to polyamory, and especially among young people, drama runs high and relationships can be of very short duration.) I have reservations about most of the well-known "guides to polyamory"—the original edition of Easton and Hardy's *The Ethical Slut* is unrealistically optimistic about the average person's boundaries and self-awareness, while Deborah Anapol's *Polyamory: The New Love Without Limits* is distastefully critical of monogamous people.[53] As polyamorous people continue to think and write about their practices, however, they are creating a body of literature that normalizes bold honesty, self-responsibility, and explicit negotiation in romantic and sexual relationships. Some of the more recent guides to polyamory reflect this maturing perspective; for instance, Tristan Taormino's *Opening Up* is honest about the emotional and social pitfalls of a polyamorous life and also draws parallels between polyamorous and BDSM negotiation techniques.[54] Nearly all writing on polyamory is frank about the importance of sexual health, however, and reflects a high level of real-life responsibility within established polyamorous communities. Most polyamorous people consider it appropriate to be tested for STDs every six months to a year, to use condoms and other latex barriers, and to ask new partners about their sexual history and request that they also be tested. This kind of forthrightness about sexual health provides a level of safety that many serial monogamists forego, due to negative judgments around using barriers or being tested routinely. Whereas to many mainstream people, using a latex glove or a dental dam for manual or oral sex is tantamount to accusing the other person of being diseased, in polyamorous circles, such safer sex practices are well within the realm of normal and are often explicitly encouraged.

The emotional necessities of polyamory can also prepare one to deal with emotionally challenging situations elsewhere in life. A polyamorous friend, now functionally monogamous due to new motherhood, told me that having been polyamorous helped her and her husband with adjusting to their new baby. Although it is not often spoken of, monogamous mates sometimes feel jealous or left out when a new baby arrives and soaks up all their spouse's time, attention, and energy. The lull in the couple's sex life that is common after a birth can be hard on new parents, and without well-established patterns of honest communication about sexuality, sexual frustrations can add to the stress of learning to care for a child. Referring to the practice of some poly people where one or more partners are "primary" while others are emotionally, financially, or otherwise "secondary" in priority, my friend told me that she had begun to think of the baby as the "primary" for both herself and her husband, while their relationship with each other was temporarily "secondary." Since she was already accustomed to relationships rising and falling in priority over time, this model gave my friend a way to deal with the new emotional structure of her family. It also provided a framework for her and her husband to negotiate new boundaries and discuss feelings of jealousy or isolation.

My friend's story highlights the fact that polyamory, while most often spoken about as a sexual practice, is actually a relational approach. No one would dispute that multiple healthy relationships with many friends and loved ones are possible: one can love a spouse, children, parents, and friends without necessarily denying any of those individuals the attention and care they deserve (limited, of course, by the number of hours there are in a day). Polyamorists differ from their monogamous neighbors in their belief that more than one of those relationships can be intimate, sensual, and/or sexual. (Among some monogamous couples I have known, *any* deep emotional intimacy outside of the marriage is considered suspect, such that even the close nonsexual friendships pursued by some polyamorous people are considered inappropriate "emotional affairs." Accordingly, some polyamorous people consider themselves to be poly even when they only have one sexual partner.) Affirming that love grows when shared, polyamorists strive to move fluidly between their many relationships, being as present as possible for each one, but acknowledging that every individual involved has many responsibilities—to family, to work, to self, to the divine.

It's true that this strategy sometimes fails, just as it does for monogamous people: job responsibilities can interfere with obligations to one's partner or family, a love affair can pull energy away from family or work, and overwhelming family and job demands can make romance and sex all but impossible. In their quest for ever more love and beauty, polyamorists sometimes bite off more than they can chew, and all their relationships suffer. The open-and-honest rhetoric of polyamory can also sometimes mask genuinely addictive behavior: a constant pursuit of New Relationship Energy at the expense of loved ones, work responsibilities, or health. For the most part, however, balancing complex priorities is simply part of being human. Every human being is part of a web of relationships and responsibilities; to love another truly means accepting her as an independent being with her own agency, not an object that can be isolated and possessed.

The gift of polyamory is the shameless embrace of that multitudinous web, and further, the conviction that contentment is not enough when joy is possible. Practical and emotional reasons make monogamy a fully satisfying and joyful choice for many people. Yet the polyamorous insistence that emotional intimacy and connection (both sexual and merely sensual) need not be restricted to a single partner resonates with the needs of our isolated, touch-deprived society.

Adolescent Sexuality

I have argued for a pluralistic erotic ethics on two bases: (1) to continue to destabilize dualisms and hierarchies that contribute to power inequalities, exploitation, and the violation of civil liberties in our society, and (2) to make a place at the table for the special insights that sexual minorities have to offer to the mainstream. In order to emphasize how the concerns and experiences of sexual minorities resonate throughout society, however, I want to call attention to a group of people whose sexuality is at least as scrutinized, demonized, legislated, and pathologized as that of BDSM practitioners: American adolescents.

Regardless of whether we became sexually active during our adolescent years or waited until legal adulthood, most of us probably recall adolescence as a time of intense erotic curiosity and no small anxiety. Adolescents frequently have powerful sexual drives but half-formed personal boundaries. Accordingly, it can be difficult for young people to discern what kind of erotic contact is comfortable and when to say no. For some adolescents, anxieties about their and others' boundaries leads to overcautiousness and abstinence that they later regret; for others, overindulgence or risky sex are what leave their scars. Still others manage to have sexual encounters that they later look back on with fond nostalgia. Because of adolescents' relative inexperience and the intense physical and hormonal changes of puberty, young people's processes of discovering themselves sexually are never going to be smooth and untroubled. It is clear, however, that the conflicting and sometimes nigh-hysterical messages that American adolescents receive from their culture make this process of discovery far more fraught than necessary. In 2009, *Good Morning America*'s Claire Shipman interviewed Sharlene Azam about her documentary and book on oral sex among American teenagers, *The New Goodnight Kiss.*[55] According to Azam, about 54 percent of girls between the ages of fifteen and nineteen were engaging in oral sex. *Good Morning America* presented the prevalence of this behavior as "shocking" and clearly damaging, though its negative consequences were not specified. The girls' report that oral sex was sometimes traded for relationship favors was condemned as "prostitution" without addressing the similar commodification of sexual favors within adult long-term relationships (recall the presentation of sex as a "wifely duty" in marriage manuals of the past). While the commodification of sex in love relationships is problematic regardless of the age of the participants, the notion that this kind of "prostitution" is a special perversion of naïve adolescents is clearly inaccurate.

In the conservative Culture and Media Institute's web coverage of the spot, one of the girls is quoted as attributing adolescents' oral sex to peer pressure and the desire to be in a relationship—again, pressures and concerns that are equally valid for adults.[56] The article goes on to blame ABC and other television networks for encouraging teens to have sex by portraying sexual relationships in network programs. I was struck, however, by the maturity and frankness of the girls interviewed in the brief spot. Significantly, the remark of one girl that girls "have urges that need to be taken care of" is entirely ignored by the C&M coverage; within the *Good Morning America* spot, the comment is contradicted as naïve, as an adult "expert" responds, "They believe they can hook up the way guys do and not care, but unfortunately, they do care." Here, the reality of girls' sexual desire is denied, while the sexist assumption that young men are never hurt by attempts at "no strings attached" sex is reaffirmed. The *Good Morning America* spot closes with GMA regular Robin Shipman urging parents to police their adolescent girls' (*not* their boys') sexual behavior. She quotes the opinion of parenting "experts" that if parents do not "engage constantly now with their children, they will be at risk"—though at risk for what, again, is not explicitly stated.

This news spot and accompanying coverage highlights much of what is toxic about American society's attitudes toward adolescent sexuality. Despite girls' increasing willingness to claim their sexual desires, their statements that they want sex are ignored, while comments about social pressures are assumed to be "the real story." Adolescent boys are portrayed, in contrast, as being emotionally unaffected by sexual encounters. There is no acknowledgment that women and men both desire sex and romance, but are frequently conflicted about those desires in culturally inflected ways. Throughout the spot, teen sexuality is presented as a crisis that requires the intervention of parenting experts and emergency response tactics on the part of parents.

The failure of abstinence-only sex education has been well publicized. Recent studies demonstrate that comprehensive sex education programs are more effective than abstinence-only ones at encouraging adolescents to delay sexual activity, and they also result in higher rates of condom and contraception use.[57] The refusal of American society to acknowledge the mutuality of young men's and young women's sexual desire, however, is one more way in which adolescent sexuality is pathologized. Deborah Tolman's research on sexuality and desire in adolescent girls examines what Tolman calls the "dilemmas of desire," the paradoxes teenage girls face as they seek to express their sexual desire while avoiding the potential negative consequences of doing so. In addition to practical risks such as pregnancy and disease, Tolman argues, girls must deal with this cultural lie about gender and desire:

> [. . . C]oncerns about boys are glossed over by the assumption that adolescent boys not only are sexual beings but are overwhelmed by their sexuality, and that such intense sexual desire is *a natural and normal part of male adolescence and male sexuality.* A gendered perspective on adolescent sexuality offers more explanation for what is behind the urgency of resisting girls' sexual desire: Girls' *lack* of desire serves as the necessary linchpin in how adolescent sexuality is organized and managed.[58]

This construction of desire as being natural to boys and unnatural to girls has made the entire subject of girls' sexual desire a taboo topic in our culture, while paradoxically, female adolescent bodies are used to sell products in the media. Sexualized images of adolescent girls are constructed to be desired rather than identified with, however; they do not effectively support real-life girls in articulating their own sexuality. Many of the girls Tolman interviewed lacked language to describe their desire; some were unable to claim it at all, describing intercourse as something that "just happens" without their agency, responsibility, or conscious choice. Those who did feel strong sexual desire, and especially those who chose to explore it outside monogamous relationships, were faced with the destructive effects of acquiring "bad reputations" or being considered "sluts." Tolman writes:

> But "it just happened" is much more than a story told by yet another girl to describe her individual experience. [. . .] It leaves out the ways in which girls are under systematic pressure not to feel, know, or act on their sexual desire. It covers up both our consistent refusal to offer girls any guidance for acknowledging, negotiating, and integrating their own sexual desire and the consequences of our refusal: sexual intercourse—most often unprotected, that "just happens" to girls.[59]

The adolescent girl on the *Good Morning America* spot who is willing to claim her own urges—and the general frankness of that interviewed group of girls—suggests that in the years since Tolman did her study, there are at least some areas of the United States in which girls can openly claim their desires and not risk destroying their social reputations.

Consent and the Law

Yet if this change in adolescent culture around sex is indeed occurring, it seems to be doing so as an act of open rebellion against long-standing cultural attitudes. In *Erotic Innocence*, James R. Kincaid argues that Americans' fears around adolescent sexuality are partially due to a conflation of adolescence and childhood, as well as to an unhealthy sexualizing of children. He notes that childlike features are often considered attractive on adult celebrities, and that other problematic blurrings between childhood and young adulthood take place, as in some legal cases where eighteen-year-olds are legally defined as children. For example, Kincaid discusses the 1992 case of a female high school teacher accused of having sex with a male student. He describes how, in the course of the trial:

> A smart and active older adolescent is shrunk into a child, a generic "essence-of-child," by this cultural story, remolded as passive, innocent, and guileless. His actual age, activities, particularities are melted away to fit our needs. Alan's sexual activity in particular is fashioned as unwilling, forced onto him or drawn from him "unnaturally." [60]

Such a portrayal of the young man would have seemed ridiculous if his sex partner had been an adolescent girl, not an adult—and, in fact, still seems so to me. A sexual relationship between a high school teacher and a student is clearly inappropriate. Yet to characterize a young man's desire for his teacher as unwilling or unnatural denies the reality of adolescent sexuality and the young man's agency in making his own choices (or mistakes).

Pat Califia and Gayle Rubin have both argued eloquently for the revision or abolition of age-of-consent laws and have done so primarily from a gay and lesbian perspective. Gay and lesbian adolescents coming of age in small towns, they point out, are unlikely to know any other gays or lesbians their age. Although age-of-consent laws are intended to protect adolescents from exploitation and abuse by adults, in practice they also criminalize loving relationships between older adolescents and lovers who may be only a few years their senior. In the late 1970s, Califia wrote extensively of the "gay kiddie porn" panic, in which some gay men were given prison sentences decades long for possessing nude photos of their seventeen-year-old lovers. She quotes an interview with then fifteen-year-old Mark Moffett, a gay adolescent who had had several cross-generational sexual relationships.

Moffett claims that (at least in the 1970s) it was more common and easier for gay boys to seek out gay adult men than vice versa, and that boys often had to take the initiative in propositioning and persuading their partners. Although Moffett is concerned with the abuse of young men by older adults, he is skeptical of the effectiveness of law in addressing the problem, stating, "I don't think rape is being stopped now with age of consent laws."[61]

Law is an ineffective tool for determining whether an individual has the ability to consent. The issue of how to determine when a person is able to give informed consent to sex is complex, and some mature and worldly fifteen-year-olds are more prepared to do so than some sheltered twenty-one-year-olds. I recall a lesbian couple in my high school social group: one girl was fifteen, the other nineteen. In Texas, their sex was legally considered statutory rape, even though the younger girl's parents were aware and approved of the relationship. "Rape" is a horrifically inaccurate label for an act between women who were young and in love. Although I hesitate to recommend the abolishment of age-of-consent laws, laws that consider the difference in age between a minor and her lover rather than setting a rigid age of consent may help to acknowledge the sexual desire of adolescents while also legally protecting them from exploitative relationships (and such laws already exist in some states). Such laws might also provide flexibility on the penalties for such crimes depending on circumstances; for instance, a nineteen-year-old convicted of statutory rape for having sex with her fifteen-year-old lover might be more appropriately required to do community service or undergo counseling, rather than serving time in prison.

Adjustment of the laws, however, will not solve the larger social problem where adolescent sexuality is denied, pathologized, or commodified at the expense of those to whom it rightfully belongs. When laws to criminalize sex involving adolescents do not exist, other laws used to persecute sexual minorities are employed instead. In the case of the female high school teacher and her student discussed above, there are clear moral and ethical issues: teachers have power over students in a way that can easily become abusive when a sexual dynamic is introduced, and teacher-student sexual relationships blur boundaries that make a classroom a safe space for students, therefore damaging the teaching relationship. There is no question that a high school teacher who chooses to have sex with a student should be fired and the student moved to a new environment. In order to make the teacher's violation a legal matter rather than simply an ethical one, however, she was charged not with statutory rape (sex with adolescent boys was not illegal in California at the time), but with "oral copulation"—in other words, performing oral sex. Though she was ultimately acquitted, the fact that this rarely-enforced law was employed to criminalize otherwise legal sexual behavior demonstrates how laws governing the consensual sexual behavior of adults continue to be enforced primarily against suspected sexual deviants. The intense medical, legal, and social attention—most of it negative—given the sexual activity of adolescents makes them functionally sexual minorities.

The Spiritual Gifts of the Erotic Adolescent

To return to the notion of the special gift of the sexual minority: adolescents are in the process of actively discovering their sexuality, and their experiences can contain a freshness, intensity, and sense of wonder that can be difficult to recapture in adulthood. Yet this period is also a delicate time, and most adolescents lack appropriate guidance about how to create the sex lives that they want. In my studies as a scholar of religion and pop culture, I fell in love with two spiritually-inflected autobiographical narratives of adolescent sexuality: *Blankets* by Craig Thompson and *Diary of a Teenage Girl* by Phoebe Gloeckner, both of which are graphic novels.[62] The novels represent two dangers of the failure to provide healthy models of adolescent sexuality. Thompson, raised in a fundamentalist Christian environment, is taught that his gentle, romantic sexuality is poisonous, but he is redeemed by a bittersweet high school romance and a religious vision that convinces him of his sexuality's sacredness. Gloeckner's alter ego Minnie, on the other hand, is the child of a negligent mother, and her quest for love and approval leads her into a variety of abusive sexual relationships, including one with her mother's boyfriend. Although both accounts contain wisdom and insight about the dilemma of adolescent sexuality, the novels have been marketed to adults, not adolescents; both authors' works have even been labeled as pornographic.[63] Yet such personal narratives can tell adults a great deal about adolescents and adolescents a great deal about themselves. The glorious intensity and novelty of adolescent desire could be celebrated as a gift to be cultivated and carefully given, rather than repressed, condemned, pathologized, or denied.

In "An Immodest Proposal," Heather Corinna describes what many Americans think of as the "ideal" first sexual intercourse narrative: in the context of a long-term loving relationship, an adolescent boy gently coaxes an adolescent girl into safer sex, and the experience (while not blissful) is intimate and brings them closer together. Corinna points out that this narrative is the best many young people hope for, but it is so much less than what is possible. Instead, she imagines a sexual initiation narrative that recognizes young women's sexual desire and makes them central, active players in sexual activity. In a series of rhetorical questions, Corinna describes an upbringing that prepares young people (and especially young women) to embrace desire, maintain healthy boundaries, and say both yes and no forcefully and with confidence. She writes:

> What if she came to sex already comfortable with her own body and sexual response, and her male partner had the expectation not of being the person who *taught* her about sexuality, gave it to her, or took it from her, but rather of *learning* about it with her? [. . .] What if she were reared with the absolute that women experience, initiate, and pursue desire, and that it is completely acceptable to do so with great enthusiasm?[64]

Corinna imagines a paradigm shift around sexual rhetoric, especially for young people, but ultimately for everyone: a shift that focuses on pleasure, respect, and mutual satisfaction. Some young people are already being raised with these values. The Unitarian Universalist curriculum *Our Whole Lives*, for example, is a sex-positive course in health, decision-making, intimacy, and sexuality that provides age-appropriate materials from the kindergarten to the high school level. Although American culture is a long way from being able to offer effective sex education in public schools, progressive sex education is available in some religious communities and could be offered even more widely. In our schools, even a move away from ineffective abstinence-only education to courses that emphasize safer sex and clearly given, enthusiastic consent could make a positive impact on how adolescents encounter their own sexuality. Religious communities also have opportunities to offer meaningful rites of passage for adolescents—rites that convey new responsibilities and also communicate positive erotic values. In the Pagan community, many women's groups celebrate first menses rites, and a smaller number celebrate manhood rites for young men; some organizations, such as the New Hampshire-based Temple of Witchcraft, are also in the process of creating rites of passage specifically for queer adolescents.

Embracing Erotic Diversity

I hope it is clear that by emphasizing the importance of marginalized sexual perspectives and recommending "a preferential option for the queer," I am not proposing a kind of queer spiritual essentialism. Queer-identified people, transpeople, BDSM practitioners, adolescents, and other sexual minorities do not have exclusive and proprietary access to certain kinds of divine knowledge. A marginalized social position creates an *opportunity* for subversive thought and boundary-breaking experience; it does not guarantee it. Further, the broad "queer perspective" as described by Heyward and others is clearly one that can be held by heterosexuals with entirely mainstream sexual tastes—they simply have not been driven towards such a perspective through social disapproval or persecution. The special insights of queer perspectives are not solely available to the sexual minority communities—nor are they applicable only to those communities. Kate Bornstein affirms this perspective when she writes: "I think anyone who regularly walks along a forbidden boundary or border (gay/straight, sober/drunk, female/male, black/white, etc.) has the potential to gain some degree of spiritual awareness."[65] The experience of seeing both sublime and destructive possibilities that lie outside of the narrow realm of social respectability is open to all people; sexual minorities are simply pushed toward these extremes of experience by persecution.

Sedgwick and Althaus-Reid's definitions of queerness imply that to be queer means never to be fully mainstream, always to be just a bit subversive, uncomfortable, disruptive. I want to reiterate, therefore, that making a commitment to tolerating and even including marginalized perspectives in social discourse about sexuality does not mean that alternative sexual practices will or should become common. As our taboos shift and change, our sense of what is boundary-crossing and subversive will shift, and formerly mainstream practices may drift to the margin. A sufficiently pluralistic erotic ethics, however, guarantees that sexual minorities' rights will be protected and that they will be part of our society's conversations about the erotic. There is a sickness at the heart of our culture that denies the body, denies the necessity of intimacy and of pleasure. Perhaps some of the solutions to that sickness will come from the innovation that takes place at the fringe.

Embrace of erotic diversity is one such solution. Our society desperately needs a pluralistic erotic ethics—a system of belief that celebrates the diversity of consensual erotic expression and affirms a variety of erotic values and approaches. When we pathologize sexual variation rather than locating our social problems in the exploitation, abuse, or commodification of bodies, we fail to uphold pleasure as a sacred human birthright. If we seek to judge loving, consensual contact, we restrict our freedom and deprive ourselves of the diverse insights and pleasures cultivated in alternative erotic communities.

In *The Book of the Law*, the Goddess is "divided for Love's sake"—split into Self and Other so that desire, discovery, union, and communion will become possible. Yet in the myth retold in Chapter 1, the fragmentation of God Herself does not end with a single division. S/he divides again and again, giving birth to a series of gods and spirits and ultimately, the earth and all its creatures. This myth is not one that values homogeneity. Instead, it is a story of desire driven by difference. Variety is at the heart of an erotic cosmology: without difference—without the revealing mirror of the Other—comparison is impossible and our self-exploration severely hampered. The intimate exploration of difference is part of the self-discovery of divinity, coming to know all its parts through the loving touch of our many hands. Deviance, then, is potentially divine.

Chapter Summary

Chapter 3 continues on the subject of social justice by arguing for a pluralistic erotic ethics similar to the systems of virtue ethics already employed in many Pagan communities. I follow Gayle Rubin in seeing a notion of benign erotic variation as necessary for encouraging healthy, consensual erotic expression and protecting the civil rights of all people. Today, rigid value hierarchies around what constitutes "good sex" and "bad sex" continue to shape public policy, with the result that some marginalized groups have used the demonization of even more marginalized groups to gain political capital (as anti-porn feminists did by attacking lesbian-feminist and other BDSM practitioners). Such sexual panics have been used to pass legislation enforcing mainstream or conservative gender and sexual norms. Rubin argues instead that all forms of sexuality have moral complexity and can be engaged in healthy or abusive ways.

Following the model set by liberation theologians, I suggest that because of their marginalized position, communities of sexual minorities have unique insights into spirituality and society—in this case, specifically into issues around sexuality. Here, I include BDSM practitioners, transgender people, and polyamorists as sexual minorities that are vocal in the Pagan community.

In Chapter 2, I argued that BDSM practitioners have a negotiated, consensual, dynamic approach to hierarchical relationships that could inform all kinds of sexual power dynamics. In this chapter, I also examine BDSM as a mystical or shamanic practice that may help practitioners and their communities to challenge oppressive social hierarchies and values. The transgender community is presented as providing a unique perspective on the biological and cultural components of gender and on the destructive qualities of our society's still-rigid gender-based binaries and hierarchies. Polyamorists, in turn, offer models for alternative family structures and relational interdependence that may address isolation arising from the isolated nuclear family. Finally, I consider adolescents as a pathologized sexual minority whose infantilization helps to prevent the establishment of effective, publicly-funded education around sexual ethics and health, as well as depriving adults of adolescents' valuable commentary on coming of age in our current sexual culture.

CHAPTER 4

The Sacrament of Touch

You cannot keep your hands from loving what they have really felt.
Jacque Lusseyran, *And There Was Light*, p. 28

Following Ellison's thought that the erotic has been unhelpfully privatized in Western culture, I began this book by framing it in the widest terms that I could: a cosmology where eroticism is a primordial power that sets the universe itself into motion and knits together the diversity of Being. The erotic is not something confined to our bedrooms or to our most intimate relationships. In fact, our failure to acknowledge the erotic flow of life force in our larger social structures and daily lives contributes to economic and social inequalities as well as to poor individual and social health. Having explored the erotic in cosmic terms and then as a driving force for social justice, however, we can turn to the erotic as we experience it most readily: as a quality of touch shared between individuals.

Sexuality was a major focus in my discussion of society and the erotic, as sexuality remains a challenging and emotionally charged topic, particularly for Americans. Sexual touch, however, is only one type of erotic touch, and it is one that is appropriate in relatively limited circumstances. Sexual energy is rarely (if ever) proper in the workplace or the classroom, though many of us spend the majority of our waking hours in these locations. In contrast, bodywork offers models for intimate but explicitly nonsexual touch that can inform the interactions of our daily lives. For many bodyworkers, massage is an explicitly spiritual practice that brings therapists into close contact with the physical and emotional wounds of their clients. Yet to so personally encounter human suffering on a daily basis is not necessarily the hardest part of the job. For some, the challenge of giving pleasure in a container that feels safe to both client and therapist—without allowing simple pleasure to blur into the sexual—is far greater. The ability of skilled bodyworkers to provide erotic pleasure in a nonsexual container provides a model for mindful touch as a spiritual or religious practice.

Mindful Touch as a Sacrament

Our lives are ripe with opportunities for meaningful contact of many kinds. The erotic flows when we meet a store clerk's eye and smile; when a handshake offered to a colleague communicates genuine warmth; or when a hug to greet a friend is given with deliberate affection, not as a perfunctory gesture. Our awareness of and openness to the eroticism present in all our relationships can enrich them in subtle yet powerful ways. Some of us—particularly those raised in physically expressive families—may find it easy and natural to take advantage of these opportunities; it may already feel normal and appropriate to exchange affectionate touch with friends and family. Others, however, grew up where physical affection was less routine—sincere but awkward, betraying discomfort with the body; limited to children or members of the opposite sex; or lacking entirely. For the touch-deprived, the combination of clearly defined boundaries and openness to contact modeled by bodyworkers may help us to welcome more touch into our lives.

What I intend, however, is more radical than the idea that increased levels of touch will result in greater health and happiness. *Loving, consensual touch can be a deliberate religious practice.* When we experience life force as fundamentally driven by erotic connection, a desire—perhaps even a holy obligation—arises to facilitate such connection between and among groups, between individuals, and within ourselves. A practice of consensual, pleasurable touch—offered formally, as in a session of bodywork or in a ritual of healing, or informally, as in the warm squeeze of a friend's hand—encourages empathy and connection and opens the way to the love of God Herself, reflected back at us when we look deeply into another's eyes. We are each the eyes and hands and lips of the gods, and in touching each other, we experience our infinite variety and particularity, as well as the underlying life force that joins us and makes us one.

I think of this practice—the practice of making contact as both an ordinary and a profoundly holy act—as *sacramental touch.* Here I wish to gently and respectfully pry the word "sacrament" out of the specific context of Christian churches, yet retain some of the connotations of the term.[1] "Sacrament" has been used historically within Christian traditions to refer to acts where the special presence of the divine can be felt: "an outward and visible sign of an inward and invisible Grace," as the Anglican Book of Common Prayer puts it.[2] Theologians have differed widely over the particulars of "grace," so I will make what I think is a fair generalization: grace is divine love offered without the loved one having done anything to deserve it. Some theologians think of it as being given in spite of the recipient's wrongdoing.[3] Yet grace is also the term for the love poured out on infants—born a mix of darkness and light, with the moral potential to do both right and wrong. So, to translate this theological terminology into words more familiar to a non-Christian audience: a sacrament is an act that demonstrates the presence of unconditional divine love in the world. While "love" and "the erotic" are not things we can see or lay our hands on, touch is their visible and tangible sign. "Sacrament" captures the joyful reverence and deliberate intentionality that a deeper understanding of the erotic can bring to the touch we offer.[4]

A Brief History of Sacramental Bodywork and Healing Touch

The idea of sacramental touch draws on a long cross-cultural history of healing. In the West, the recorded practice of healing through touch goes back to at least 1553 BCE (in the form of the Ebers Papyrus). Other recorded accounts can be found in the Bible, in the medical writings of Greek physician Hippocrates and the Roman physician Galen, and throughout the history of Christianity.[5] Eastern cultures also feature a variety of touch therapies; in both Chinese medicine and Indian Ayurvedic medicine, various points on the body are stimulated in order to balance its vital forces. These systems have similarities with the ancient Greek theory of bodily humors, which the technique of anatripsis ("rubbings") was meant to regulate. In some cultures, techniques similar to what we now think of as massage were used to dispel possession by disease-causing harmful spirits.[6] As history of massage researcher Robert Calvert describes, the traditional systems of medicine described above (and those of Australia, Africa, and other regions) arose as part of integrated cultural worldviews that included the spiritual.[7] To put the matter into modern terminology, these traditional systems of healing touch are based on specific theologies of the body and its relation to the cosmos.

Contemporary Paganism's theologies of the body have been specifically impacted by alternative therapeutic movements of the nineteenth and twentieth centuries. As religion scholar Catherine Albanese describes, many of these movements were expressions of religious or spiritual belief systems: physical religions in which healing and curing were "the central ritual enterprise."[8] In both Europe and the United States, such therapies were influenced by the philosophy of mesmerism, formulated by Franz Anton Mesmer in the late eighteenth century. According to Mesmer, living beings possess a vital force—a life energy or etheric fluid—that can be manipulated through the mesmerist's intentional gaze, gestures, or touch. Appropriate manipulation of this force (which Mesmer called "animal magnetism") was thought to have the power to encourage healing and restore health.[9]

Mesmer was not the only thinker to assert that all living things share a vital life energy.[10] This therapeutic theory came to be more generally called "vitalism," and it resonated with other alternative religious movements influential in the nineteenth century, such as Swedenborgianism (based on the theology of eighteenth-century philosopher and scientist Emanuel Swedenborg). In a model that Albanese describes as "conflat[ing] spirit and matter," Swedenborg portrayed the natural world as created by the emanations of an "infinite fountain" of divinity[11]—a view not unlike that of the American Transcendentalists, some of whom perceived nature as both a tangible sign of divinity and actually infused with the divinity it represented.[12] From the idea that a divine life energy is immanent in nature came alternative therapies that viewed the cause of disease as divergence from the body's natural state. Such therapies included the natural food movement; homeopathy, which treated disease through a theory of "natural harmonies" between causes and cures; and osteopathy and chiropractic, which sought to restore the free flow of the body's energies through muscular or skeletal manipulations.[13]

As the contemporary Pagan movement grew rapidly in the 1970s, many were directly influenced by the work of twentieth-century psychoanalyst Wilhelm Reich, whose name for the energy pervading the natural world was "orgone." Reich believed that orgone could be transmitted from being to being through touch and that an enormous amount of orgone was built up during sexual activity, then released during orgasm. If the proper flow of orgone was blocked because of emotional or other dysfunction, taught Reich, mental and physical disease would result.[14] Reich's notion that pleasure is essential for health, along with his hostility toward authoritarianism and patriarchy (and his ultimate martyrdom in prison), made him a popular figure among readers and writers of *Green Egg*, the most influential Pagan magazine of the 1970s. Reich was seen as a fellow nature mystic (though not a theist, as historian Chas Clifton points out[15]), and his theories provided semi-scientific support for Pagan experiences of divine energy immanent in the body and in sexuality.

In contemporary Paganism, Reich's psychoanalytic approach also mixed with existing beliefs about sexuality and spiritual liberation drawn from nineteenth- and twentieth-century Western esotericism. The practice of sexual ritual, beginning in the Victorian era, sought to resolve the pathologies created by repressing human sexual instincts (as described by Sigmund Freud and others[16]); many practitioners embraced vitalistic theories similar to Mesmer's. Techniques for sexual ritual were drawn from Western esoteric sources and psychoanalysis as well as from Eastern tantric traditions. As historian Hugh Urban argues, in the nineteenth and twentieth centuries, those who embraced the practice of sex magic did so in search of personal spiritual freedom, but also because they believed that sexual liberation would contribute to a new social order free of racial and gender-based oppression.[17] These social goals were widely shared by mid- to late-twentieth century Pagans, many of whom drew on Western esoteric traditions as direct sources for the development of new Pagan rituals and practices.

Such egalitarian goals also motivated the practice of "naturism" or nudism. Naturists understood nudity as a state that erased signs of social rank, encouraged physical health, and emphasized personal and social freedom. Accordingly, Wiccan publicizer (and likely founder) Gerald Gardner incorporated the practice of ritual nudity into his budding religion in the 1950s.[18] Particularly in Wicca, but also in a variety of Pagan groups newly emerging in the United States, Pagans came to see acts of consensual sensuality and sexuality as moments where human beings actively channel divinity: the "acts of love and pleasure" named as the Goddess' rituals in the Wiccan Charge of the Goddess.[19] Although not all the practices above were understood as directly therapeutic in the same way as chiropractic or orgone therapy, they were nevertheless conceptualized as healing practices—behaviors that sought to restore the natural harmonies of the body with nature and the cosmos and so encourage personal and social health.

Contemporary Paganism today includes a number of different practices that could fall under the umbrella of sacramental touch. Sexual ritual for the purpose of communing with divine energies (often performed privately) continues in Wiccan and other traditions. Additionally, Pagan and other spiritually-oriented sex workers offer sexual healing services to those traumatized by sexual assault or other abuse, or who are simply struggling with body shame or social anxiety. As mentioned in Chapter 1, some of these sex workers think of themselves as "sacred prostitutes" and understand their work as continuous with ancient temple prostitution (an idea that, according to recent scholarship, is probably based on a scholarly myth[20]). Such sexual healing work can take a variety of forms, from counseling, to hands-on sexual instruction, to performance art. Its goals include helping individuals cultivate pride in their bodies; gain a sense of ownership over their sexual power; liberate repressed sexual energy; learn how to communicate about sex and maintain healthy boundaries; and practice techniques to achieve ecstatic or mystical states, as well as simply to have more enjoyable sex.[21]

Pagans engage in a variety of sensual and therapeutic forms of sacramental touch as part of religious practice. Ethnographer Sarah Pike relates how, when she was first beginning to study the Pagan community, a priestess offered to teach her about ritual by designing a special one for Pike's thirtieth birthday. In an emotionally intimate encounter, Pike was lovingly bathed and anointed with oil by the priestess in sacred space, then invited to encounter the divine in the priestess' eyes and in a mirror. Pike was deeply moved, and she describes the ritual as "an opening between self and other, between worshipper and goddess, between researcher and informant"—one in which the sensual elements were a key part of the powerful experience.[22] Such a ritual uses touch as a method to bless, heal, and create awareness of divine presence. Similar uses of touch can be found in the Five Fold Kiss, a Wiccan practice in which words of blessing and kisses are bestowed on various parts of the recipient's body, as well as the practice of sacred group massage, in which a recipient receives gentle, nonsexual massage from the members of the group while all participants verbally affirm her beauty and holiness. In groups where I have observed or experienced this kind of ritual touch, a common blessing on the recipient is "Thou art Goddess" or "Thou art God." The blessings affirm the divine presence within the recipient's physical form, and in some rituals, those initiating touch are also characterized as "the hands of the gods" or as otherwise being possessed by divine energy. Such rituals may be performed simply as blessings, or as practices meant to dispel shame or encourage healing from trauma.

Many Pagans also practice modern forms of laying on hands meant to heal injury or disease. In some Pagan communities (as in both amateur and professional bodywork circles), it is common to be trained in Reiki, an originally Japanese spiritual touch practice that encourages physical and emotional healing through improved flow of life force.[23] Reiki practitioners and those seeking healing sometimes participate in "Reiki shares," free-of-charge gatherings where the group lays healing hands on each participant in turn. Some practitioners also offer personal, one-on-one Reiki sessions for a fee. Such practices affirm widespread Pagan beliefs that mental, physical, and sexual health result from the proper movement of natural and/or divine forces, and that appropriate, intentional touch helps to correct and strengthen that flow.[24]

Seeking and Expressing the Divine through Touch

Sacramental touch is an attempt to raise intentional touch to an art form, and through that intentionality, to foster healing and divine connection. Because of her clinical background, a professional therapist is capable of touch that is highly intentional: she facilitates change in the client's body by combining knowledge of anatomy, physiology, and technique with an emotional tone that is appropriate to the client's condition. Yet the study of biology is not necessary to bring reverence or a sense of the holy to touch. While an amateur may not be able to provide the same degree of technical specificity as a professional therapist, it is definitely possible to bring a similar level of attention—and as bodyworkers know well, the healing effects of bodywork do not come solely from physical manipulation. One of the most transformative bodywork experiences I've had came from an amateur. Some time after becoming a licensed massage therapist, I attended a massage party where therapists had been invited to give lessons and demonstrate simple massage techniques to anyone who wanted to learn. I was offered a Thai yoga massage by a man who had had perhaps a dozen or so hours of informal lessons. Although I had received more technically proficient bodywork before, the level of attention and sensitivity I experienced in that massage brought me not only a delightful awareness of my body, but induced a light state of trance. I have rarely, if ever, been more lovingly touched. The generosity shown me by the man I met at the event—a total stranger to me—gave me a new benchmark for how mindfully I could approach not only professional bodywork, but also casual touch among friends.

I opened Chapter 2 by recalling how massage school helped me to fully accept the oneness of the human body and spirit. This unity is one reason that bodywork can be so intensely intimate. Bodywork educator David Lauterstein is fond of saying that when a practitioner lays her hands on a client, she is laying her hands on an entire history: all the experiences of an entire life, and all the inheritances of that person's ancestors. Deane Juhan puts it similarly, but emphasizes individuals' active role in shaping their own bodies: "[O]ur flesh is the written record of our lives, faithful in every last detail. We are all to a large degree self-made; we are both the hand and the clay."[25] Because body and spirit are one, to touch another's body is also to make contact with another *soul*—an act that deserves nothing less than awe. When we touch each other, we help to shape the paths of each other's lives: we guide each other toward or away from health, cause pain or pleasure, nurture connection or break it.

In practicing sacramental touch, the development of greater health, pleasure, and connection are paramount. Touch is a tool that we use to better understand each other and the divine. Deane Juhan argues that intentional touch such as that used in bodywork can help to re-educate the body out of restricted, painful movement habits and create healthier, more pleasurable patterns. But this re-education does not stop on the physical level. Because all our perceptions are filtered through the body, touch is an essential component of engaging with the universe and coming to a deep knowledge of the divine and ourselves. In his bodywork manual *Job's Body*, Juhan engages the biblical book of Job, in which Job tosses aside the advice of others—his friends, his wife, and religious experts—to directly question God about his life's unjust tragedies. Job's confrontation with God gives him an experience of the vastness and incomprehensibility of the divine, but he also experiences his oneness with it. "[I]n my very flesh do I see God," Job declares (19:26). Juhan responds:

> For Job, this was revelation—the perception that God was in his very flesh, in the throbbing of his heart, in the singing of his nerves, in the coiling of his muscles, to be touched and felt more intimately than an embrace. Not eternal judge and Justice, but immanent event and consequence. Not perfect and complete, but growing and changing, absorbing each new development and getting every apparent contradiction its day. Not infinite in material fact, but infinite in possibilities. Most important of all, not remote and imperturbable, but present and responsive; not unapproachable, but in need of an active partner.[26]

As mentioned in the Introduction, in Western theology, the notion that the divine is immanent in the world (including the human body) and is in a constant state of growth, change, and development is a feature of process theology. Rather than being transcendent—outside of us or the universe—God changes with us, struggles with us, *is* us. Juhan alludes to this conception of the divine as being in process, and he suggests further that partnership, or *relationship*, is an essential feature of the way the divine expresses itself—in the universe in general, and in human beings in particular. By striving with God, Job finds God in himself in a concretely physical way.

In his essay series "The Seven Dimensions of Touch," bodywork educator David Lauterstein also meditates on the parallels between divine–human and human–human relationships. Stressing the importance of the bodyworker's intention when first making physical contact with a client, he writes,

> The first element or dimension of touch occurs when we lay our hands upon someone. Just laying down a hand establishes a point or area of contact. [. . .] This single dimension of basic contact sums up everything so eloquently depicted in Michelangelo's famous Sistine Chapel painting, where God and Adam reach toward each other. The power of this picture is that it depicts with stunning visual beauty the creation of life.

> Creation occurs in the moment and in the space of touching another. This is the relationship.[27]

For Lauterstein, when contact occurs—between the divine and the human, or between two human beings—a relationship is initiated that will affect and change both parties. The intention and emotional quality of the touch are incredibly important, as they set the tone for the immediate interaction and perhaps for the entire relationship. As Lauterstein puts it, "If where we first make contact [during bodywork] becomes habitual and not the result of a sacred decision, then we've squandered an incredible opportunity."[28]

According to Lauterstein, conscious touch provides the opportunity to establish an "I–Thou" relationship. This term comes from the work of Jewish philosopher Martin Buber, who defined such a relationship as one where neither party is objectified, but where each is fully and lovingly acknowledged in all of his or her complexity.[29] For Buber, "I–Thou" describes the ideal relationship between a human being and the divine. The qualities of such a relationship—compassion, openness, caring, respect, commitment—form a potential basis for all other relationships. The divine is seen and cherished in all things and especially in other people. Whether or not we are professional bodyworkers, making first contact—particularly when touching another physically, but also when first making eye contact—can provide an opportunity for holy connection. A relationship is forming, or continuing; the tone for that day's interaction is being set, and every new moment of first contact is one in which we can consciously touch another person in the knowledge that he is also God Herself. A massage therapist who makes initial contact with a client hurriedly, in a careless touch that communicates distraction and stress, may lose that client. When we touch others without empathy, without the awareness that we are each children of the divine, opportunities for joy and pleasure are lost; we are more likely to find ourselves feeling lonely and disconnected.

To see others as divine, however, is not to neglect ourselves. Our family members, our friends, and the strangers walking down the street may all be part of God Herself, but each of us is divine as well, and worthy of love, care, and protection. I am the assigned caretaker of my own body; while I may seek to treat others with love and reverence, it is most important that I maintain an I–Thou relationship with *myself.* I may wish to approach every relationship and every first touch knowing that I and they are both divine, but I also have the obligation to set good boundaries with those who will not treat me with respect, caring, or even basic politeness. Alone at night on a dark city street, I do not smile and make eye contact with strangers; I know myself to be in a relatively vulnerable position, and I value my safety enough that I choose not to call attention to that vulnerability. My potential mugger may be as much a part of God Herself as I am, but he is unaware of that fact and means me harm. Those who are unwilling or unable to engage in relationships of mutual respect deserve compassion, but not indulgence. Sacramental touch happens most easily in a place where both parties feel safe, and while there may be exceptions to this principle—as I will discuss later in this book—I do not wish to recommend thoughtless risk-taking.

Touch Deprivation

In a touch-deprived society, there is a certain amount of risk involved in initiating touch at all, as most Americans are socialized to touch very little. In *The Power of Touch*, Phyllis Davis reviews a cross-cultural study of how frequently friends sharing a meal touch each other in an hour, on average. Puerto Ricans touch around 180 times an hour, French friends 110 times, and American a mere two times.[30] This relative lack of touch is distressing when we consider how central touch is for physical and mental health. Deane Juhan summarizes the research of Rene Spitz and others on infant deaths in orphanages in the late nineteenth and early twentieth centuries, which in some American cities could range from 90–99 percent. At the time, the condition was called *marasmus*, meaning "wasting away." Those babies who did survive past the first year of life were severely physically retarded, a condition now referred to as "deprivation dwarfism." At length, the cause was determined to be a combination of understaffing and neglect. The psychological wisdom of the time advised not "overhandling" babies, lest they become spoiled; additionally, staff generally only had time to clean and feed the babies, not to handle them or play with them. When extra staff were added at such orphanages to cuddle the infants, however, mortality rates dropped dramatically. Babies who were already experiencing retardation showed improvements in weight, height, energy, and cognitive functioning.[31] Anthropologist Ashley Montagu observes that, during the period where medical experts considered touch inessential for young children, marasmus "was found to occur quite often among babies in the 'best' homes, hospitals, and institutions, among babies apparently receiving the 'best' and most careful physical attention."[32] Without touch, it seems, no medical intervention was sufficient to nurse a sickly infant back to health; frequently-handled babies seemed to inexplicably thrive in impoverished and unhygienic conditions while the isolated children of the upper class sickened and died.

Animal studies have demonstrated that tactile stimulation is a biological requirement for infant mammals of all kinds, not just for human infants.

For many species, a baby animal must be stimulated in the abdominal and perineal areas—usually by the mother's tongue—or it does not begin to urinate or defecate and soon dies of toxicity. Though this information is common knowledge among those who raise animals, it is not widely known in the scientific community or the general public. For example, Juhan describes an experiment where researchers attempted to raise animals in a germ-free environment. Nearly all the animals died of distended bladders until, on the advice of a former zoo-keeper, they began to stroke the newborns after each feeding.[33] Without tactile stimulation, it seems, certain autonomic functions cannot be established, and growth does not proceed normally.

In *Touching: The Human Significance of the Skin*, Montagu reviews a number of animal studies that explore the impact of touch on a variety of young animals. In one study of baby monkeys, the young animals were provided with a cloth-covered structure that radiated heat to simulate the presence of a mother, as well as with a similar structure built solely from wire mesh. The wire mothers were equipped with milk feeders. To the surprise of the research team, the young monkeys spent more time clinging to the cloth mothers than to the wire mothers who provided their food: the comfort of the heat and softness created attachment behaviors much more strongly than the availability of milk.[34] Pleasurable touch, researchers concluded, was the key element that caused the young monkeys to form an attachment with their caregiver. Further, monkeys who were isolated when young developed a variety of hyperactive, apathetic, and violent behaviors that appeared much less frequently in monkeys raised by their mothers. Isolated monkeys showed a greater frequency of rocking and other self-stimulating behavior, suggesting the effects of deprivation, and isolated monkeys who later had offspring showed abusive behaviors toward their young with much greater frequency than their socialized counterparts.[35]

In other studies, rats that had been handled extensively in infancy showed increased immunological responses—and therefore, greater resistance to disease—that persisted into adulthood.[36] Rats who were handled frequently and affectionately not only showed calmer, less anxious, and friendlier behavior, but also survived surgical procedures to remove endocrine glands at a rate 63–66 percent greater than their less-handled siblings.[37] A third rat study showed that affectionately handled rats grew faster, showed greater problem-solving ability, and responded to stress with less anxiety.[38] Frequent nurturing touch during the developmental period, it appears, increases the ability of mammals to endure physical trauma, as well as tending to produce an even temperament.[39]

Based on a variety of animal studies and infant studies, Montagu concludes that frequent, loving touch is necessary for the physical, psychological, and social development of children. Western practices of leaving babies alone in cribs for extended periods of time or refusing to pick crying babies up for fear of spoiling them, he argues, may hinder children's development and increase undesirable behaviors. In one study, mothers of babies with colic—a condition of infants where they cry persistently, apparently in response to gastrointestinal distress—were advised to hold their babies whenever they cried and to let them feed on demand, not on a schedule. These infants showed a 70 percent decrease in crying compared to the control group. The researcher theorized that this effect was not simply due to the comforting nature of the mother's touch, but also that, as with the animal mothers licking their young, tactile stimulation helps to regulate the function of the digestive tract.[40] Other studies suggest that increasing physical contact between mothers and hospitalized, premature babies helps to speed weight gain and central nervous system functioning in the infant, while also hastening the mother's recovery from childbirth.[41] Frequent nurturing touch, it seems, can potentially benefit the health of both parent and child. Accordingly, Montagu speaks admiringly of cultures in which infants and very young children are held or worn in slings almost perpetually by caregivers, and families commonly share a bed.[42]

Some researchers have suggested that depriving individuals of touch can have a wide-ranging social impact. For example, cross-cultural studies by developmental neuropsychologist James Prescott correlate low levels of affection shown toward infants with violent behavior in adults. Prescott's studies claimed that whether or not newborns were carried on the body of a caretaker during the day correlated with the relative levels of violent behavior in that society to an accuracy of 80 percent. Societies that did not punish premarital sex also showed a lower frequency of violent behavior than those that did.[43] Davis emphasizes a telling point in Prescott's research: in societies where infants were not frequently touched, but premarital sex was not punished, there was still a lower incidence of rape, murder, and theft than in more touch-averse societies. If early touch deprivation does indeed cause antisocial behavior in adulthood, this finding suggests that the effects can still be ameliorated by affectionate touch in adolescence, and perhaps even later in life.[44] Accordingly, some researchers have focused on touch therapies as a treatment for behaviorial problems. For example, studies conducted by researcher Tiffany Field suggest that increasing the amount of safe, boundaried touch that disturbed adolescents receive may help to relieve symptoms of mental illness. In one study where adolescents institutionalized for psychiatric disorders received massage for half an hour every day for a week, their levels of stress hormones (cortisol and norepinephrine) dropped, their sleep patterns became more regular, and incidents of inappropriate sexual and other behavior decreased, leading to earlier discharges from the psychiatric program.[45]

Given the evidence that touch is necessary for human development, the increasingly draconian laws around touch in American public schools—where children spend the majority of their days—do not bode well for children's health. Although adults are understandably concerned about the potential for inappropriate touch, touch deprivation may be one of the culprits behind increased rates of hyperactivity in children, as well as a contributing factor in anorexia and other eating disorders in adolescents and adults.[46] Yet educators fear being fired or taken to court for affectionate behavior towards students. Rules prohibiting hugging and handholding between students are also increasingly common in junior high and high schools.[47] Such rules, while well-intentioned, may be helping to create the kinds of abusive behavior that they seek to prevent. For example, in one study of touch behaviors in groups, Tiffany Field observed adolescents socializing at restaurants in France and the United States. French adolescents leaned on each other, rubbed each other's backs, and put their arms around each other with much greater frequency than American adolescents. American adolescents, on the other hand, showed much higher rates of self-touching—twirling hair, cracking knuckles, playing with jewelry—as well as higher rates of aggressive behaviors such as pushing and hitting. (Pushing and hitting, notably, may actually be socially acceptable expressions of desire for closeness, but they may nevertheless segue into genuinely violent behavior.) Field suggests that the affectionate touch among the French adolescents may attenuate aggressive urges and behaviors, and that the high-touch patterns found throughout French society may correlate with its relatively low levels of adult violence.[48] By forbidding affectionate touch between American students, school authorities may be creating social situations that encourage bullying, gang violence, and abuse in dating relationships.

Distinguishing Qualities of Touch

In part, American culture polices touch so rigidly because the boundaries between sexual and merely affectionate touch have become unnecessarily blurred. There are complex historical reasons for this situation which I cannot adequately cover here. Current trends can be traced at least to the Victorian era, in which restricted and stylized forms of touch became one marker of membership in the upper class.[49] This cultural history has contributed to widespread touch deprivation, but also to a more subtle problem. Today, many individuals have not learned to make clear distinctions about different qualities of touch; they may assume that all touch is sexual, or fail to recognize sexual touch when it occurs.

This confusion is due to a lack of experience with different qualities of touch. Because of the potent biological drives and reproductive necessities surrounding sex, nearly all cultures condone sexual touch in some form. Sexual touch may only be allowed in specific contexts and between specific classes of people, however, and some sex acts may be classified as taboo.[50] When highly restricted sexual touch is one of the few permitted forms of touch available, individuals may develop a tendency to perceive *all* touch as potentially sexual. If one's wife or husband is the only acceptable source of touch in one's life, the romantic and sexual nature of that relationship may spill over uncomfortably, or erotically, when touching others either by chance or through a deliberate act of transgression. Further, when sexual needs are not being met due to cultural restrictions and inhibitions, sexual frustrations may color other kinds of physical encounters, creating a sexual charge where a sexually-satisfied individual might experience only comfort or affection.

Phyllis K. Davis describes Americans' confusion of loving touch with sexual touch as a system of pathological thinking that she calls the Phobia Syndrome. This set of social fears hinders individuals from getting the kinds of touch they want and need. For example, people may withhold affectionate touch out of a fear of being condemned as promiscuous or homosexual. Among and between families, adults may restrict their touch behaviors so as not to accidentally encourage adultery or incest, or to avoid accusations of child molestation. Additionally, the cultural tendency to sexualize large areas of the body—often everything but the hands, arms, shoulders, and head—may cause any touch to be inaccurately read as sexual.[51]

Davis states that "in order to learn to differentiate between hands that touch to turn on, and hands that touch to convey comfort and affection, [individuals] must experience both."[52] In her practice as a somatic psychologist, she has found that most individuals find it relatively easy to make distinctions between kinds of touch when they are taught directly. In group workshops, she asks participants to choose partners for a touch exercise in which one partner touches the other while focusing one by one on the emotions of tenderness, anger, detachment, and sexual arousal. The receiver attempts to identify each touch without being told which emotion is being expressed. Having removed the expectation that every touch is potentially sexual, most individuals are able to begin making clear distinctions between these types of touch either immediately, or with a few repetitions.[53]

Deliberately developing the ability to distinguish types of touch—and being clear with our own intentions when we touch another person—helps to ensure that both persons in the exchange will have a positive experience. It is common in our culture for both men and women to initiate sexual touch out of a desire for affection. This tendency, when unconscious, can result in mysteriously unsatisfying sex and a nagging feeling of disconnection. When both individuals in a relationship have a larger menu of kinds of touch to choose from—and particularly when sexuality has not been muddled with merely affectionate touch—they will be more able to identify their desires and negotiate a physical encounter that satisfies them both.

Bodywork may be one mechanism through which individuals can learn to make such distinctions. As Juhan writes, the explicitly healing, nonsexual context of bodywork creates a safe container for clients with abuse histories to experience and identify caring touch.

> [Emotional and sexual abuse] deeply confuse pleasure-giving touch and nurturing human contact with unwanted intrusion, and by conditioning us to avoid intimate contacts of all kinds and at all costs this intrusion compels us to starve ourselves of the necessary ingredients for our survival. Bodywork is often a dramatic solution to these crippling confusions and reactions, because it gives an individual a safe place and a safe time in which to work out meaningful distinctions that have been hopelessly scrambled.[54]

For those in recovery from abuse, and even just for individuals for whom affection and sexuality have become too thoroughly conflated, sessions with a reputable bodyworker may help to reset expectations around touch and boundaries. Massage is a physically intimate activity, but one that is performed with strict boundaries by professionals with training in ethics. Full-body massage generally takes place with the client fully unclothed and covered with a drape. The bodyworker uncovers one part of the body at a time as he or she works, while the genitals remain modestly concealed. Rarely, breast massage may be performed at the special request of the client. In bodywork, partial nudity is not sexualized, but is understood as a helpful practice for healing—the skin can be treated with moisturizing oils, and the lack of constraining layers of clothing assists the practitioner in detecting and addressing patterns of tension in the muscles. To the degree of the client's comfort (usually determined in an interview before the session), the bodyworker may manipulate parts of the body that have not been touched nonsexually since the client's childhood. The thighs and gluteals—large muscles involved in walking that frequently carry tension and may contribute to hip or back pain—may be massaged, or the abdomen, the domain of the powerful hip flexor muscles and muscles involved in digestion. Such work opens opportunities for clients to experience these sometimes highly emotionally charged parts of their bodies in new and positive ways.

Cultivating a Spiritual Practice of Touch

As a professional bodyworker, and as a frequent and enthusiastic recipient of professional bodywork, I feel there is much to be gained from bodywork sessions, especially for an individual who wants to take up the practice of sacramental touch in her life. Bodywork can help an individual to re-establish a healthy relationship with her own body, as well as improve physical and emotional health. Many clients see bodyworkers to relieve chronic pain and tension caused by repetitive physical activities or stress. However, bodywork can also reduce anxiety and improve sleep; correct postural misalignments that contribute to chronic pain; improve circulation, speeding up the flushing of cell wastes and bringing nutrient-rich blood to cells for growth or repair; and support immunological functioning by interrupting the highly stressed, fight-or-flight state in which many of us spend our workdays.[55] An individual who is healthy and pain-free has much more to offer emotionally and physically to others. Further, experiencing the touch of someone who has practiced healing touch for years can communicate the qualities of such touching much more effectively than reading this entire book. I have often been amazed at how potent the effect of an experienced bodyworker can be. The touch may be as simple as resting a hand on the back of my head, but I may feel tension begin to leave my shoulders, my pulse and breathing slow, my jaw unclench. A skilled bodyworker brings more than anatomical knowledge and technique to a session—he communicates a sense of peace, relaxation, and contentment that cannot be learned from a book.

Like Davis, however, I think that it is both possible and desirable for those who have no interest in touching others professionally to cultivate a powerfully healing touch. To declare that every touch can be a holy expression of God Herself reaching out in love is a first step towards the practice of sacramental touch; to be able to *feel* this truth emotionally, and begin to see the divine reflected in another's eyes, is a second. But once the mind and the emotions have committed to the practice, the body must get involved, and here a bit of study and practice will not go amiss. Some of us have been blessed to be raised in families who were easy and open with touch, and may already know, for example, how to put an arm around a friend in a way that feels comforting and warm rather than restrictive. But for those of us who are dealing with a lifetime of feeling awkward about touching, books and classes on massage and other kinds of healing touch may provide tools, not just to give a good backrub or quiet a screaming infant, but also to achieve particular qualities of touch. Davis, for example, relates an anecdote about an overly enthusiastic hugger in one of her workshops, a large and cheerful man whose hug knocked out her breath and popped one of her ribs. No doubt the man wanted to convey warmth and a lack of sexual intent, but leaving Davis winded was probably not the effect he had intended![56] If we approach touch as a religious practice, it must be taken seriously as a *practice*—something that we commit to, study, repeat, and refine. A commitment not just to touch, but to touch others *well*, is an important part of sacralizing touch in our lives.

I do not, however, wish to pass on the anxieties that characterized my own struggle with touch. My personal inhibitions around touching others revolved around a fear of touching them awkwardly, uncomfortably, at the wrong time, or in the wrong way. Often, these inhibitions persisted even when I knew that reaching out would be welcome. Wishing to change my own attitudes towards touch, I actively sought out opportunities to give or receive backrubs, hold a friend's hand, or cuddle while watching a movie. Still, as a young person, I turned down perhaps as many opportunities as I embraced, due to a fear of not doing it *right*. Kind and affectionate friends and lovers helped me develop confidence in touching others, and I began to offer touch to those who were struggling as I had. Formal training in bodywork freed me further. I learned to speak to clients about sensitive issues around touch, as well as create a context where the expectations around touch were clear. The opportunity to practice pleasurable and therapeutic techniques of touch in a safe environment added to my confidence.

Today, I am probably still more reticent with a hug than my friends for whom near-constant affectionate touch was a part of their growing up. However, the tools that bodywork training gave me have made me far more comfortable with touching others. When I urge readers to cultivate an educated touch, I do not mean that if touch cannot be done well, it should be avoided. Rather, I hope that greater intentionality, knowledge, and consciousness around touch will help readers overcome shyness or awkwardness. It will also help to ensure that the touch that they initiate will be received warmly and with pleasure. Putting the too-forceful bear hug of Davis' anecdote aside, no training is necessary to give a hug or hold a hand. While books and classes will help make us highly skilled practitioners of sacramental touch, there is no substitute for touching others as our hearts move us.

Honoring Boundaries

In touch, spontaneity is best mixed with empathy, connectedness, and respect. If we seek to touch others as a sacrament—to touch them in a way that initiates an I–Thou relationship between divine selves—it is important to honor others' boundaries and make sure we have their consent to touch them. As someone who has historically been anxious about imposing on others, I have benefited greatly from the explicit consent involved in giving or receiving bodywork. A professional bodyworker will interview the client before the first session and explain how the session will proceed. Clients have the opportunity at that time to set certain body parts or certain kinds of touch as "off-limits" for whatever reason. Whether giving or receiving bodywork, I find myself much more comfortable and confident after this direct exchange of expectations.

As mentioned previously, some sexual minorities have created subcultures where explicit negotiations around touch are the norm. Books on kink occasionally even provide lists of activities that potential partners can fill out and exchange, indicating their likes, dislikes, limits, and potential interests. Given the wide range of intense activities that they engage in, most BDSM practitioners feel that keeping play "safe, sane, and consensual" must involve a great deal of extremely explicit communication. This kind of negotiation is also common in the polyamorous community, where feminist sexual ethics have been a powerful influence. Many polyamorous people are aware that, due to gender-related socialization patterns in our culture, both men and women frequently struggle to say no to sexual advances they do not want. It is clear that this problem must be addressed from both sides: the person who is approached must develop the assertiveness needed to say yes or no with confidence, while the person who is approaching should seek clear consent and be emotionally prepared to be rejected. At polyamorous gatherings, newcomers are often encouraged to ask verbally before touching, especially when approaching someone they do not know. The ability to ask, "Would you like a back rub?" and receive the answer gracefully is considered a sign of simple good manners, while a flirty "May I kiss you?" can often come off as romantic. In either of these cases, to ask and be refused should still feel better to both parties than to touch without permission. No one likes to feel a new acquaintance stiffen under their hands, or to have a love interest flinch back from an attempted kiss.

Such verbal conventions also help to ensure (though by no means guarantee) that neither party engages in sex he or she does not want. Even sensitive and caring people can misread a partner's body language when in the throes of passion, especially if the partner has mixed emotions rather than being entirely unwilling. It is a painful thing to learn that a partner feels guilty, embarrassed, or uncomfortable after what the other person thought was a pleasurable night; it is even worse to find oneself in a pattern of silence, repeatedly having sex that one does not want in an attempt to avoid hurting the other person's feelings. When frequent verbal communication is part of a relationship from the beginning, conversations about sex become normal and expected, and the highly-charged silence that can arise around unsatisfying sexual encounters is less likely to form.

The art of gaining verbal consent, once mastered, is rarely inappropriate. With a little practice, the question "Would you like a hug?" can come to seem natural. In certain situations, however, it may also be necessary to assess consent nonverbally. Based on the context in which I am meeting someone for the first time, I may open my arms to offer a hug. Usually the offer is accepted when the other person opens their arms in return. Occasionally, however, I am offered a handshake instead—a clear signal about that person's boundaries. Such boundary-setting is not a personal rejection. When another person has stricter boundaries around touch than I do, most likely those boundaries have nothing to do with me as a person, and much to do with how he understands appropriate behavior (with new acquaintances, with people of a certain gender, with non-family members, etc.). By shaking a new acquaintance's hand instead of insisting on a hug, I honor who he is and how he is coming to the relationship that is about to form. While a greeting hug may seem friendly and natural to some, to others it seems overly familiar and invasive; trust will be better established by maximizing the warmth and genuineness of the offered handshake, not by pushing the boundaries of what the other person finds comfortable. When assessing consent nonverbally with someone new, it is important to watch the other person's body language and facial expression and aim for the lowest common denominator—the type of touch that both parties signal they are comfortable with. The intimacy of touch can always be increased later, when there has been opportunity for more communication. When first meeting, it is most important that the touch that is exchanged be genuine and comfortable. If the first contact sets the tone for a relationship, than an invasive or boundary-pushing first touch can leave a negative impression.

That being said, there are times when taking risks around touch is worthwhile. For instance, I have casually reached out and touched the arm of a new acquaintance, then watched her eyes and body language to determine whether more touch will be welcome. A flinch, breaking of eye contact, or turning away of the body is a definite no! Increased eye contact or a touch in return could be read as an invitation for more, while a sudden increase in flirtatiousness might indicate that the other person interprets friendly touch as sexual—a sign that, if my interest is purely friendly, I may want to make my motivations clear. Preferring to err on the side of caution, if I cannot read the other person's reaction to a quick, casual touch, I will frequently ask for consent verbally before trying again (perhaps asking for a goodbye hug). While one may wish to be extra-conservative around taking such risks in a professional environment or when interacting with someone with a different cultural background, the chances of a brief, clearly nonsexual touch being perceived as a violation are relatively low when one is in a familiar social context.

I myself have sometimes been a person who is more open to touch than I initially seem. For example, I have misjudged the formality of a situation and offered my hand while my new acquaintance simultaneously opened his arms for a hug. I usually laugh off the moment of awkwardness and accept the hug—a situation where I end up accepting a more intimate touch than I initially signaled I wanted. I see no problem with adjusting in the moment this way, although obviously, no one should ever insist on a hug, or in fact on any kind of touch at all. If the other person hugs me only out of a sense of social expectation, the result is generally an unsatisfying, perfunctory touch. If we seek to touch as a religious practice, we may wish to increase the amount of touch in ours and others' lives, as well as its intimacy. It is important, however, that we do so in a way that increases our connection with others rather than alienating them. Physical contact is not something to be pushed on others, no matter how "good for them" it may be. A sacramental touch is always extended as an invitation, not as a demand.

When Lauterstein speaks of an "I–Thou" relationship between bodyworker and client, he has in mind a relationship where the bodyworker is aware of the client as a whole person, not just as a bundle of muscle and bone to be manipulated. Nor is the client a mere consumer of therapy, whose opinions about her own body should be waved aside if the bodyworker believes she "knows better." Such approaches are objectifying—literally, they treat a human being like an object by refusing to take that person's subjective experiences seriously. Every person's experience of her own body is valid; we are all equally part of God Herself, and when we speak of our own experience, those words need to be taken as seriously as though they came from the mouth of God. This is not to say that we are all equally empowered to make medical diagnoses; Western medical science is designed to treat illness based on objectively observable signs and symptoms, not on the report of the patient, and this approach has validity. Rather, it is the person's report of their own emotions, sensations, and experiences that must be taken seriously. Though a person who is convinced that she has cancer or a psychological disorder, without a formal diagnosis, might need to be challenged, a person who reports pain or distress should be believed and treated accordingly. Each of us has the right to express what gives us joy or pain, what makes us feel comfortable or distressed, and to have those boundaries honored by others.

The act of honoring boundaries may be frustrating at times. Others sometimes make decisions about touch that seem unhealthy to us—by isolating themselves, engaging in extreme physical activities, or seeking sex in risky situations. When interacting with loved ones, it may sometimes be our place to persistently offer loving touch to an isolated relative, or to tell a careless friend that we are concerned for their health and safety. In some cases, when our needs for touch do not match those of a loved one—such as in a sexual relationship where two partners have extremely different or otherwise incompatible desires—to truly honor the other person as a "Thou" rather than an objectified "It" may cause the relationship in its current form to end. If I try to maintain a relationship by asking the other person to change or deny his desires, my refusal to acknowledge those desires as holy objectifies him. Knowing that I cannot healthily fulfill my partner's desire for touch, it honors his divinity to give him freedom to explore it—either by changing the form of the relationship, or by ending it, depending on my own boundaries and needs.

Embracing One's Own Right to Pleasure

In the act of honoring others' boundaries and needs, we must not allow our own desires to be constricted by objectifying relationships. Part of acknowledging our right to pleasure means refusing to accept poor treatment. Yet some kinds of poor treatment are subtle: not abuse, but parasitism. To remain close to a person who seems intent on self-destruction is always difficult. I applaud those who can remain healthy and sane while connected to partners in the throes of addiction, or who comfort and shelter friends who will not leave their abusers. In some cases, however, these unbalanced relationships threaten to drain the helper's own emotional and physical reserves. In seeking to care for others, I have ignored my own needs and treated myself as less important than the person I was caring for. Hoping for love or affection from my struggling loved one, I have denied myself the kinds of touch I desired and pushed my body to a state of exhaustion. But I too deserve to be a Thou, not an It.

For some people, the true challenge of seeking out I–Thou relationships is not the challenge of honoring others' desires, but of prioritizing their own desires equally. There are some exceptions to this principle—for example, the parents of young children inevitably sacrifice their needs to their children's some of the time. Still, an exhausted and emotionally drained caregiver is a less effective caregiver; he cannot be fully present for a child in the way a happy, rested caregiver can. When we model self-care by seeking out assistance and taking time for ourselves, we model healthy patterns of emotional independence (and interdependence) for our children. In honoring others as embodiments of the divine, it is essential that we do not forget to honor ourselves equally—particularly those of us who are caring for children or elderly family members, or who work in helping professions. While the desire to help, comfort, and heal is a holy one, I do not honor the part of God Herself that is me when I neglect my own well-being. My self—the piece of consciousness that I think of as "Christine"—is the designated caretaker of my body, my personality, my heart, and my soul. If I do not care deeply about my own thriving, I am throwing away the gift of my embodiment. No one else is in a better position than I am to discover and give me what I need. How can I genuinely turn to another and affirm "Thou art God/dess" if I cannot see God/dess in myself?

Although receiving professional bodywork is an excellent method of self-care, amateur classes or working through a massage book with a friend can be equally effective. Delightfully, taking classes in massage usually provides opportunities to receive touch, as one's classmates need bodies on which to practice. When the amount of touch that one receives is substantially increased, the personality changes that result can be dramatic and surprising: I myself noticed a major reduction in anxiety and elevation in mood during my massage training, during which I received at least one massage a week.

For those whose busy lives do not allow time and space for taking classes, Davis' *The Power of Touch* includes a number of exercises designed to increase one's comfort level with touch, and which can be done at any time with a friend or loved one.[57] Some exercises include reflecting on one's personal history of experiences with touch, beginning with childhood; observing and describing one's current touch habits and behaviors; initiating small, casual touches with strangers to observe their reactions (and one's own); exchanging backrubs or foot massages with friends; experimenting with different kinds of sensation with a loved one (scratching, rubbing, stroking, tickling) and asking for feedback; asking for a hug when in a familiar group; cuddling with a partner with the lights on and off and observing the differences; and much more. Above all, Davis encourages her readers to seek out others who are willing to experiment with and talk openly about physical contact. These allies can provide essential support in transforming one's own relationship with touch.

Certain volunteer opportunities can also challenge and expand one's own touch boundaries. Hospitals sometimes need volunteers to cuddle or massage newborns, and homes for the elderly often welcome volunteers to spend time with their residents. In my work as a massage therapist, I have visited assisted living centers to give neck and shoulder massages to the elderly. In many cases, the fact of my touch was just as important to my lonely clients as the reduction in muscle tension. One woman consistently told me that the visit was the highlight of her week. Yet one need not be a trained professional to hold an elderly person's hand or give a gentle shoulder rub. The elderly are a particularly touch-deprived population, especially those who are unpartnered. Family visits may be rare, so that the only touch an elderly person receives is the clinical one of a doctor or nurse. When we offer touch to those who are deeply in need, we open to the profound healing capacities of ordinary touch and of our own capacity to channel divine love.

A Dance of Self and Other

In the story of God Herself creating the universe, erotic touch is a creative force, and also a method of discovery. By dividing in two, S/he is able to experience parts of hirself as though they were Other. This process of dividing and reconnecting is also a process of coming to self-knowledge, a way to reintegrate the separated parts with deeper understanding. We are pieces of God Herself interacting with each other, and touch is a vehicle for discovering the nature of ourselves and others.

In *Job's Body*, Juhan's argument is similar, but more nuanced. Although he sees the potential for divine self-knowledge in each one of us, such knowledge must be cultivated; it is not inborn. He believes that Western culture, with its tendency to treat mind and body as separate rather than parts of a whole, encourages a state of dissociation that leads us to treat our bodies as separate entities. Yet we are embodied beings; the body is our only vehicle through which to experience the world. Bodywork, Juhan believes, provides a partial solution: by increasing awareness of the body, bodywork can re-educate the individual into perceiving all her parts as a whole. Such a perception makes the individual more fully capable of experiencing herself, the environment, and the divine. He writes:

> A successful relationship with reality must be learned as a dance is learned—not by contemplation but by participation, by an exertion of will, by movement, action, and response. In order to derive strength from God and the angels, it is necessary to enter into active contact with them, to push against them with our muscles and feel them with our senses. We are formed not by abstract laws but by the intimacies of a wrestling match, where we struggle with universal forces until we begin to feel our individual forces grow in relation to them.[58]

Juhan's claim is simple—there is no contact with or knowledge of the divine without the body, no growth or evolution except through flesh. We learn where our selves begin and end by making contact with others, defining the "I" and the "Thou"—but I would add that in making that contact, we also experience our underlying sameness, the pulsing current of life force that flows through all things. We are many and also one, magnificently beautiful in our particularity; yet in our moments of unity, our deepest ecstasy can be found. For love's sake we have been divided; and in touch, in erotic expression, a circuit is completed for love to flow.

Touch others, then, as a sacrament, a holy rite in which divine love is tangibly felt; for your hands are the hands of the gods; your eyes full of compassion or yearning are theirs, and from your loved ones' eyes they look back; your kisses of affection or lust will be pressed on the Goddess's own lips. "*For I have been with you from the beginning, and I am that which is attained at the end of desire.*"

Chapter Summary

Chapter 4 explores mindful, intentional touch as a sacramental practice, one that can offer experiences of the presence and unconditional love of the divine. Spiritual approaches to bodywork provide a framework in which people can cultivate the erotic in their personal lives. Such a practice of sacramental touch can include the sexual, but our society's need for safe, consensual, affectionate nonsexual touch is in many ways more urgent.

I begin with a brief history of sacramental bodywork and healing touch, with an emphasis on the immediate precursors to sacramental touch practices in contemporary Paganism. Pagan theologies of touch and the body are strongly influenced by vitalism, the idea that all living things share a universal life energy. This belief is connected to Pagans' experience of the Goddess or the gods as immanent in nature and the body. Returning again to the idea that spirit and body are fully integrated, I engage spiritual writing on bodywork to suggest that when we touch another human being, we contact not just flesh, but the entire history of that person's consciousness. Such touching deserves reverence and intentionality.

Next, I look at studies of touch deprivation to argue that without frequent, pleasurable touch, human beings suffer severe health consequences. Studies have shown that touch is necessary for the healthy physical and cognitive development of all mammals, and that societies in which affectionate touch is frequent have lower levels of violent crime. Formal massage treatment has also been shown to speed recovery in emotionally disturbed adolescents and may be helpful in recovery from physical or sexual abuse. Using standards of well-boundaried, consensual touch derived from bodywork and sexual minority communities, as well as the practices of psychologist and touch advocate Phyllis Davis, I suggest methods by which non-professionals might cultivate healthy touch practices among their families, loved ones, and communities.

CHAPTER 5

A Love Stronger than Fear

And there came a leper to him, beseeching him, and kneeling down to him, and saying unto him, If thou wilt, thou canst make me clean. And Jesus, moved with compassion, put forth his hand, and touched him.

Mark 1:40–41

[As a sacred prostitute,] I work with people who have been raped, who are victims of incest, and who are very frightened of their bodies. I hurt for them because I know how beautiful it is to be inside your body and really enjoy it.

D'vora, *People of the Earth: The New Pagans Speak Out*, p. 145

In writing this theology, I have imagined a diverse readership: not just Pagans or sexual minorities, but all kinds of body-honoring religious people, united by a desire to perform the sacrament of touch. Each reader is a potential minister of that sacrament, a priestess or priest of the erotic divine. It is an office that can take many forms, and one with no hierarchy of merit. The volunteer who visits nursing homes in order to hold the hands of the elderly does work that is as important as that of the professional bodyworker or sacred sexuality educator. In whatever way we are each able to offer (or train others to offer) sacramental touch, we alleviate or prevent just a bit more suffering and spread the conviction that we are all holy and infinitely loved.

For myself—and here I do not automatically include the reader, as this is a matter between each person and her experience of divinity—to offer sacramental touch is not merely a calling, but also an obligation. Despite knowing how touch has transformed my life and deepened my spirituality,

I still at times struggle to offer compassionate touch when I know it would be welcomed, or find myself retreating into a comfortable bubble of personal space even when doing so will be emotionally isolating. To help those around me towards erotic health, I must constantly confront how I turn away from the demands that pleasure and ecstasy place on me.

Dedicating oneself to the sacrament of touch can be uncomfortable or even dangerous. Those acting on an ethic of the erotic may encounter HIV-infected blood while rendering emergency aid; experience job loss or damage to their reputations for hugging a child not their own; contract a painful STD from a loved one; or endure fear and emotional distress to offer comforting touch to the dying. We may be persecuted as deviants in our communities (and, sadly, living in a sex-radical community does not guarantee immunity from persecution). Shaped by our culture's unhealthy commodification of sex, we may make errors in judgment and hurt or violate those we love. Or—particularly those of us who have been physically abused—we may simply push ourselves too far and too fast, then find ourselves on the therapy couch trying to unravel destabilizing emotional backlash. Some people are called to positions of radical erotic service: surrogate partners help clients with intimacy problems in the context of sex therapy, while sacred sex workers and dominants offer themselves (legally or otherwise) to those who have not otherwise been able to meet their erotic needs.[1] Still others are called to much more subtle transgressions of Western erotophobia. All those who dedicate themselves to the sacrament of touch, however, become obligated to take risks in the name of joy and pleasure.

Too many of the early publications of the sex-positive movement assert that a simple shift in attitude will create the sex life of one's fantasies. It is not enough to declare that "Sex is fun and pleasure is good for you!" Progressives are well aware that the lingering remnants of mainstream Victorian sexual mores no longer serve us; we know that the proverbial Emperor wears no clothes. Yet a naked Emperor, contrary to the fairy tale, is still Emperor until he is dethroned and replaced. Our outdated sexual ethics need to be replaced with a value system that is equally sophisticated and compelling.

Erotic Activism on the Front Lines

Initially published in 1997, Easton and Hardy's *The Ethical Slut* was a groundbreaking work for the sex-positive movement. Its first edition, however, failed to address how being unusually open to erotic connection can make a person (especially a young person) vulnerable to predatory partners and social rejection. Nor did the book adequately discuss the good boundaries and self-respect that must be built before one is able to confidently consent to and refuse erotic activity. Even today, young women who acquire a reputation for being "easy" can become targets for sexual predators who will not acknowledge a refusal from a self-declared "slut." A friend who had worked as a stripper told me of the night she was gang-raped by a group of customers after leaving her workplace with them. As the men were violating her over her protests, one asked, "Can I fuck you in the ass?" She refused vehemently, and he refrained. This bizarre exchange highlights attitudes that still remain in our society around sex work and "easy" women, particularly when judgment-altering drugs or alcohol are involved. My friend was a stripper, and she left the building with the men; in their minds, the meaning of those actions was consent to sex, regardless of what she may have said verbally. Anal sex, however, was a special category of sex requiring separate permission, and when she verbally refused that permission, the refusal was respected.

Men and women who wish to be "ethical sluts" or activists for erotic liberation need to be aware that others in our society may see their behavior as a declaration of undiscerning sexual openness. I firmly defend the right of people to engage in sex work and to have as many sexual partners as they choose, so long as all involved are consenting and informed. But to present oneself as being open to many sexual partners can attract the attention of those who think of the human beings they desire as objects, without their own experience or subjectivity. A personal commitment to sacramental touch is no protection against those who have no context for respectful sexual contact, let alone sacred sexuality.

Certainly no amount of education in street smart, self-defense, and assertiveness training will prevent people from becoming the victims of rapists. Yet if sex-positive writers do not speak of the risks involved in being sexually open, we do our readers a disservice. As millions of survivors of rape and sexual abuse can attest, experiences of sexual assault cause wounds that can take decades to heal. I do not wish to imply that sexual assault prevention is the responsibility of potential victims: rather, sexual assault prevention is the responsibility of the entire community. Yet insofar as greater sexual openness can expose individuals to greater risk, common-sense precautions can help ensure that trauma does not cut one's career as a pleasure activist short. Like all sexually active individuals, those seeking more sexual openness in their lives need to weigh their risks and take measures to protect themselves and their loved ones: listen to their instincts when meeting new people, ask mutual acquaintances about potential new partners' reputations, or go with a trusted group when attending unfamiliar gatherings where alcohol or drugs may be involved. The ability to communicate explicitly about sex, to insist on the use of latex barriers and STD testing, and to call a halt when an encounter doesn't feel right are essential skills, as is the willingness to speak up when witnessing another being abused. Much as sexual progressives would like to live in a world where touch is considered holy, the reality is that not every partner we desire will share these values. It is possible to offer touch as a sacrament to someone who believes that sex is a pleasurable sin—and perhaps help to transform that belief in the process—but it is a necessary part of self-love to protect ourselves and those close to us while we do it.

It is true that, for sex-positive activists to be able to maintain their activism, they must carefully weigh the risks they take. Yet pleasure activism also involves confronting the community norms that make claiming erotic freedom risky—in other words, challenging the social patterns that justify sexual assault and protect sexual predators. Removing responsibility for rape from victims and placing it with the perpetrators of assault is an ongoing project of the current feminist movement (for example, in one resource from UniteWomen.org, a woman holds up a sign that reads, "I need feminism because my university teaches 'How to Avoid Getting Raped' instead of 'Don't Rape' at freshman orientation"[2]). In order to liberate themselves and others to safely explore erotic desires, sex-positive activists must create communities in which consensual erotic freedom is possible. Contributing to community conversations about consent and communication, modeling ethical negotiation practices, practicing safer sex, and holding both self and others accountable for erotic behavior can be part of a commitment to sacramental touch. For sexual minorities, subcultural erotic communities can also provide minority-focused support, protection, and education, particularly when sexual minorities are threatened with the loss of their civil rights.[3]

Eros and Mortality

Our erotic behavior should not be controlled by our fears and anxieties. Love must be the driving force behind the way we touch: love for others, love for self, love for pleasure and for the divine. In one college dorm room, the bedside poster above a colorful jar of condoms challenged, "Is your love of sex greater than your fear of death?" I've returned to this question many times when making choices around sex. More generally, the truncated question "Is your love greater than your fear?" has been a guiding principle for me when making all kinds of ethical choices. Love is profoundly risky, profoundly vulnerable-making. I am reminded of Elizabeth Stone's saying that "Making the decision to have a child is momentous. It is to decide forever to have your heart go walking around outside your body." To love deeply is to risk the agony of loss and disappointment—we cannot control our loved ones (although some of us may misguidedly try), and so we cannot fully protect them or our relationships with them from danger. Yet if we turn away from love, if we reject the possibility of pleasure and connection because of fear, we deny ourselves the joy and depth of experience that makes life worthwhile. There are things that are worse than death, and a life untouched by joy is surely one of them.

I find I cannot speak of the physical risks of erotic touch without also speaking of the emotional ones; for me, they have always been intertwined. Because of the strength of the medical model of health and sickness in the Western world, it is easy to frame emotional or moral discomfort in terms of hygiene or disease prevention. While in massage school, I lost count of the number of times that acquaintances said to me, "I could never be a massage therapist. What if I had to massage someone *gross*?" If I pressed, I generally found that "gross" meant unwashed, overweight, elderly, or sick—bodies that, in various ways, display less-than-perfect health. According to this kind of thinking, these bodies put the person who is touching in danger—supposedly in danger of disease, but perhaps more immediately, in danger of feeling disgust. The unhealthy body shows evidence of human mortality in a way that youth and health do not. I do not wish to downplay the physical risks of being a professional bodyworker—for example, like hospital staff, we are somewhat more at risk for contracting antibiotic-resistant skin infections like MRSA, as well as common flus and viruses. But this medical information is not something that my querants usually know.

Instead, what I hear in the fear of the "gross" is a reluctance to be reminded of illness, aging, and death. To come regularly into contact with normal human bodies is to know viscerally that one's own body will decline and decay. Experiential knowledge of mortality can be uncomfortable, even terrifying; facing the fact that our lives could be cut short at any time can force a sudden and painful re-evaluation of one's priorities. Unfortunately, it is commonplace to hear others react with disgust at the normal bodies of others—I recall switching off an episode of *Sex and the City* where the characters worked themselves into a frenzy of horror over the sight of a naked elderly man. Recalling that this reaction is protective, however, helps me to be compassionate. Even the wisest and most thoughtful of our spiritual leaders can find the reality of death difficult to bear, let alone those people who have not yet grappled with loss.

In my own life, health concerns have sometimes disguised lingering feelings of sexual guilt and hid the way that erotophobic cultural messages were still influencing my emotions and behavior. During a period of sexual openness and exploration, I carefully researched a number of sexually transmitted diseases to learn about risk and prevention. I recommend this project to anyone seeking to be sexually active outside of a long-term monogamous relationship, as informed consent cannot be given without good information. Yet the tone of many of the sexual health websites sent me into a tizzy of anxiety. Gone is the carefree unprotected sex of the 1970s, the brief period when the birth control pill was widely available and most known STDs could be cured with antibiotics. In addition to HIV (a virus whose associated syndrome, AIDS, is now treatable, if not curable), some formerly curable STDs such as gonorrhea have developed antibiotic-resistant strains and can be transmitted through saliva, not just through sexual fluids.[4] A diagnosis of herpes, while not at all life-threatening, can nevertheless lead to rejection by potential sexual partners, since condoms provide only imperfect protection. Barriers provide similarly weak protection from the HPV virus, some strains of which are linked with cervical and other cancers. Hysteria over this supposedly "new" virus has driven sales of a recently released commercial vaccine that helps to protect against several of the more virulent strains, but the vaccine is recommended only for women under the age of 26—up to 80 percent of sexually active adults are thought to have already been exposed.[5]

After reading up on the prevention, transmission, and treatment of STDs, one may feel that it would be easier to never have sex again. Even the faithful use of condoms, dental dams, and gloves during all contact involving the genitals cannot perfectly prevent the transmission of disease. And while websites and public health brochures often present worst-case scenarios in an attempt to frighten readers into less risky behavior, such tactics overshadow the fact that with regular testing, health education, good communication between partners, and sensible barrier use, the risk of contracting a serious disease is very low. Consider the risks associated with HPV: it is estimated that perhaps 4000 women in the United States die of cervical cancer each year. In contrast, about 40,000 Americans per year die in car accidents.[6] If we assume that most adults have had sex, and most adults have been in a car, it's clear that driving or riding in a car puts one at much greater risk. Yet in my experience and observation, it is far easier to stir up fear around STDs and permissive sexuality than it is to frighten people away from their cars.

Again, without brushing aside the very real health risks that STDs present (and the importance of testing, barrier use, and proper medical treatment), I would suggest that the perhaps well-intentioned fearmongering of public health education resources feeds our culture's underlying erotophobia. Health websites and pamphlets often present statistics and symptoms without the acknowledgment that sexual fulfillment is a component of health. In *Touched by the Goddess*, Deane Juhan writes of the way that Western culture's embrace of the medical model (with its valuable treatment of many kinds of injury and disease) has led us to forget an important principle: health is more than a lack of illness. He contrasts two paradigms as conflicting deities who can and should be complements: Asclepius, the combatant of disease, and Hygieia, the goddess of flourishing health.[7] Culturally, we have failed to honor Hygieia's holistic perspective and method of balanced, healthful habits. This failure blinds us to the fact that medical science provides tools to enhance quality of life, not precautions meant to dictate behavior. In educating oneself about sexuality, it is too easy to lose sight of *health* and instead focus solely on *disease.*

Releasing Purity Taboos

The reasons behind Western culture's historical disapproval of non-monogamy have been much analyzed. Various historians have linked the need to control women's sexuality to patrilineal inheritance practices: if property passes from father to son, then it is in the father's best interest to be sure the heir is his own offspring. Scholars of religion also point to the belief (influential in Christianity, but also present in the pre-Christian religions of the Mediterranean) that the true nature of reality is spiritual, and the body is temporary, weak, a temptation to evil, or even a prison. One might see this attitude as a theological solution to the problem of death—presented with the reality of the mortal body, the intangible spirit is elevated and declared to be eternal, rendering bodily death unimportant. Yet the impact of body-denigrating belief on sexual morality is to see sex as wallowing in the inferior physical plane. Whatever the origins of these attitudes, Westerners today have inherited a system of cultural values that frowns on having many sexual partners (though what constitutes "many," versus a "normal number," varies wildly from community to community). This disapproval of nonmonogamy also can extend to nonreproductive sex. Some groups have seized on the appearance of new STDs as a judgment on those who would violate the moral imperatives of heterosexual monogamy, as the Christian Right did with the gay community in the 1980s and 1990s.[8] Even those of us who consider ourselves to be progressive on sexual issues, however, may find ourselves using the risk of disease as a supposedly rational reason to criticize those who are more promiscuous or permissive than ourselves. "That's dangerous, s/he'll end up with a disease" can become code for "I'm not comfortable with that degree of sexual openness."

In my mid-twenties, I became aware that my values around sexuality were preventing me from having the amount or the kinds of sex that I craved physically and emotionally. I had grown up believing that sex should be spiritual and the result of a very specific and personal kind of love: a passionate connection that stems from knowing a partner well and trusting them deeply. As it turned out, however, I didn't yet have the interpersonal skills to sustain a passionate emotional connection in a long-term relationship. Because I had such high spiritual expectations for sex, I deprived myself of the everyday, undramatic sexual encounters that often help to sustain romantic connections. At the same time, I was intensely critical of friends who were willing to be sexual with people they did not know well. Despite the fact that barrier use and regular testing were the norm in my community, I described their promiscuity as irresponsible and dangerous to their health and their partners' health. Privately, I was disgusted by how impersonal it all seemed.

Gradually, I began to see how these attitudes were in conflict with my religious values. I believed—no, I *knew*—that the earth and all its creatures were brimming with divine life force. I believed that pleasure was a human birthright; I rejected the idea that pleasure that hurts no one can be wrong. I had experienced the paradoxical love of God Herself, feeling myself both cherished as an utterly unique individual, and also valued as impersonally as I might value a single cell of my body. It was possible, I realized, to treat a complete stranger as an honored Thou, not an objectified It. If I look a homeless man in the eye and smile, recognizing God Herself in him, then is it actually any harder to honor God in a new sex partner by sharing an evening of pleasure? It began to seem absurd that I thought all manner of other human interactions could be holy, but sex required a special and rarefied level of commitment and had to be held to the highest spiritual standard even to be acceptable. I hadn't realized it, but some part of me still believed that sex was dirty, sinful, and wrong; I didn't believe marriage was necessary to sanctify and transform it, but I had come up with my own set of requirements and restrictions to "clean it up"—restrictions that left me isolated and sexually unfulfilled.

I worked on the problem in meditation and ritual, with a therapist, and with self-help books. But the freeing moment came when I decided to ask a number of friends and loved ones to tell me their stories of sex with strangers—accounts of casual encounters and one-night stands. These were people I loved and respected, and I was ready to listen without judgment. Some of the stories reflected the ways our culture usually speaks of casual sex: there were random encounters without context, meaningless and so half-forgotten; there were mismatched expectations that left one partner or the other feeling rejected or used; there were accounts of awkward sex that left the participants embarrassed and eager to get away. Poor communication, lack of empathy, and an approach to sex as "getting some" rather than connecting with another person played roles in these tales. But other stories—sometimes from the same tellers—stood out in their beauty and poignancy: moments of pleasure spontaneously shared without expectation; respect and empathy offered not from a place of personal knowledge, as one offers these things to a long-term partner, but simply because both were human beings deserving of love and pleasure. *I do not have to know you to treat you with kindness and respect; I do not have to know you to show you what my body enjoys, or to offer you pleasure in return. I need only know myself.*

In relating this set of values to others, I often get a response along the lines of, "In what enlightened fantasy world does this actually happen?" The truth is, it happens already in the world we live in. It is rare, perhaps, to stumble across a stranger who communicates well and habitually treats others with love and respect; rarer still that both people in a spontaneous situation will be attracted to each other and be free to act on that attraction. But it is not a fantasy that spontaneous or casual erotic encounters can be holy. The love of the gods can move through us in any place, at any time, and when another person consciously channels that love back at us, it is a gift of immeasurable worth. Nor is the experience of this love necessarily dramatic—an exchange of kindness and affection, quietly given and quietly received, nevertheless provides us with a connectedness and well-being that colors everything we do. The notion that a supposedly debased act of "casual sex" could be among my loved ones' sweetest and most joyful memories helped me to re-evaluate the restrictions I had put on my own sexuality. Perhaps sex did not need to occur in a certain kind of relationship, or with a certain kind of spiritually elevating intention, to be holy; perhaps the holiness was within me, simply waiting for my willingness to express it.

Touch in Existing Relationships

The process of releasing touch restrictions that no longer serve us is rarely easy, and particularly when it comes to sexual touch, it can be hard on existing intimate relationships. Psychologist David Schnarch's book *Passionate Marriage* emphasizes the necessity for both people in a committed relationship to continually develop as individuals in order to maintain a strong connection with a healthy sexual component.[9] Individual growth inevitably strains the relationship's boundaries at times, says Schnarch, but if the relationship does not go through periods of struggle and challenge, it will simply stagnate. This emphasis on individuality in relationship is a paradoxical truth, but one that resonates in my own life. When the boundaries around my self have blurred in a long-term way with another's, in our fused state, neither of us is able to clearly say what we desire or to effectively negotiate how to get it.

To shift the touch dynamic in an existing relationship requires the ability to identify and express one's own desires, as well as the capacity to take either acceptance or rejection of touch as being about the other person's boundaries, not necessarily about one's own inadequacies. If I touch my partner in a new way and he flinches, it is destructive (and probably untrue) to assume that he no longer finds me attractive, or that he will never be open to that kind of touch. Instead, I can learn to talk about the kinds of touch that I want, and to ask about my partner's unexpressed desires. Being refused need not feel like the end of the world; nor should awkwardness around new kinds of touch (or around touch in general) necessarily be perceived as a relationship's death knell. Yet to make changes around touch—either to try to revive a faltering relationship, or to deepen one that is stable—can be emotionally challenging and may trigger fears about losing one's partner.

Asking for different kinds of touch in a long-term relationship may feel as though one is threatening the foundations on which the relationship is built. Indeed, at times making these requests *is* risky. For example, those who are sensitive to issues of abuse may feel intense anxiety and guilt over being asked to be more aggressive, and those who were abused as children may struggle with PTSD-like symptoms around many kinds of supposedly "normal" touch. Changing the boundaries around touching others can also be tumultuous: opening a monogamous relationship to other partners frequently brings all the existing tensions in that relationship to the fore. Powerful emotions may flow; if those involved are not able to temporarily tolerate, communicate about, and eventually resolve those emotions, the relationship may end. Additionally, not every relationship will be flexible enough to satisfy both partners' desires. The revelation of fundamentally mismatched desires, especially those that are deeply held, can sometimes bring loving but ultimately stifling romantic partnerships to a close.

If we are able to cope with the possibility of loss, however, and approach the potential exchange of touch as a sacrament that must be freely given, calm and compassionate perseverance can radically change the dynamics of a relationship over time. Scharch writes of his couples therapy practice, where honest self-expression and the ability to respect and work with a partner's different desires and experiences sometimes bring near-sexless and miserable marriages back from the brink of collapse. For Schnarch, a couple's sex life reflects their emotional life, and vice versa; the two are inextricably intertwined. In one vignette, he tells the story of "Joan" and "Bill," a couple whose relationship had become crippled by anxieties around sex. For Joan, learning to soothe her own anxiety about being inadequate and to seek her own pleasure in the female superior position—a sexual activity that was relatively new to her—also helped her husband to let go of his erection-killing anxiety around sexual performance.[10] The couple's newly relaxed attitudes allowed them to have more enjoyable and successful sexual encounters, as well as to feel more trusting and connected in daily life. Their progress, however, depended entirely on each individual's willingness to face and finally conquer long-held fears and ingrained patterns of behavior. Schnarch speaks of the practice of "holding onto oneself"—staying conscious of one's own desires while remaining present with one's partner and one's own body—as a way to deepen emotional connection and increase the pleasure of all forms of touch. Paradoxically, it is not single-minded focus on one's partner that makes this transformation possible; it is awareness of oneself, and a willingness to let go of expectations and fears around the behavior of others.

Such change is also possible in nonsexual family relationships. Phyllis Davis speaks of her dogged efforts, after the death of her father, to shift her patterns of affection with her mother. Historically, Davis' family had not been open or demonstrative about affection. In defiance of her family's norms, Davis began to tell her mother that she loved her and would hug her when arriving for and leaving after a visit. Although Davis advises not to keep touching others if they do not eventually reciprocate and especially if they actively refuse, in this case she persisted in the face of her mother's passivity, even though hugging her felt like "embracing a stiff piece of cardboard."[11] After a few months, her mother suddenly answered Davis with an "I love you" in return, and later attended a class with Davis where she was able to receive a facial massage. Within a year, Davis' mother had begun to hug Davis and other members of their family on a routine basis.

Several facets of this story strike me as essential to Davis' success. First, Davis was already in a close family relationship with her mother; her persistence would have been inappropriate with a stranger, co-worker, or casual friend. Second, her actions were an expression of the greater affection and intimacy she desired in the relationship, not a judgment on her mother's tendency to be reserved. Finally, Davis did not become upset or put pressure on her mother when she did not immediately reciprocate Davis' words of affection or hugs. Instead, Davis simply continued to offer them within her own comfort level—not giving too much and so feeling drained, but rather providing a low, steady level of affection that eventually encouraged her relationship with her mother to bloom. Davis' book *The Power of Touch* provides dozens of exercises and suggestions for small steps to increase our comfort with touch and to increase the touch we receive in our lives. Making these shifts may not be a simple process, and they may force us to face fears about personal shortcomings, negative body issues, or memories of past traumas. Some of these old, undesirable emotional patterns may never be completely eradicated. However, the process of bringing the kind of touch we desire into our lives by slowly normalizing it—as Davis did by hugging her mother—can produce lasting changes that resonate throughout all of our relationships.

Forming Alternative Erotic Communities

Within the kink and queer communities, sex and play parties have had a powerful normalizing effect on alternative sexualities. In contrast to the sensationalized mainstream view of such events as depersonalized, prurient, and sordid, many such gatherings are given as a community service and provide important opportunities for group bonding. Gatherings often require an invitation, and hosts provide an explicit set of rules about barrier use, etiquette, safewords, and how to obtain consent.[12] Patrick Califia writes of sex and play parties:

> Public sex is good for you, or at least it has made me a much more knowledgeable, kind, and skilled sexual actor. I've learned to enjoy and eroticize a wider range of body types; come to appreciate the fact that even if sex acts are repetitive, the people who are doing them are unique; and lost some of my conditioning to be afraid of a frank expression of sexual desire. [. . . I]f we are really going to understand the complexity of public sex, we must acknowledge that it is not motivated solely by lust.[13]

Califia speaks of Stuart Denton, an apparently closeted gay man who hanged himself after being arrested at an X-rated video store while engaging in mutual masturbation with other men. Regarding the incident, newspaper reporter Amy Pagnozzi argues, "We're talking about a generation [of gay men] whose very being was defined by shame, with few public places to call its own, for whom coming together isn't so easy."[14] Legal sex and play parties help to banish the guilt that drives closeted people to seek fulfillment of their desires furtively, as well as providing a safer, community-policed environment to explore those desires. Particularly in hubs of queer culture like San Francisco, sex and play parties are one venue in which sexual minorities express their values and bear witness to the acceptability of marginalized practices.

In "Playing with Sacred Fire: Building Erotic Communities," interdisciplinary sexuality scholar Loraine Hutchins frames these erotic communities for counselors and psychotherapists who are treating sexual minorities. A longtime participant-observer of sexual minority communities, Hutchins reports that beginning in the mid-1980s, there has been an attitudinal shift toward merging sexuality and spirituality, perhaps stemming from community grief over AIDS and AIDS-related deaths. Hutchins provides information about some of the larger erotic communities in the United States and introduces their norms and practices. Among the advantages of erotic communities, she lists sex education (including skills for communication, safer sex, and sexual competence); opportunities for group memorialization of those who have died from AIDS or been killed in hate crimes; permission and safe space to gain sexual confidence and explore individual sexual tastes, leading to self-discovery and healing; community safeguards against sexual abuse and predation; and opportunities for spiritual experience and group bonding.[15]

Some of the impact of sex and play parties can be obtained from any clothing-optional space, such as outdoor Pagan festivals, nude beaches, public bathhouses, or gym saunas. The consumption of mainstream media exposes us constantly to the uncovered (and frequently airbrushed or surgically altered) bodies of the most conventionally attractive 1 percent of human beings. Makeup artists, cosmetic surgeons, personal trainers, and other aesthetic professionals collaborate extensively to alter and maintain the bodies of celebrities. To compare one's own body to the products of this process—the photographs and films that feature these partially artificial human forms—is to find oneself badly wanting. Women especially seem to bond over sharing their discomfort with and dislike of their bodies: dieting, plucking, tanning, girdling, and all the while fretting over the elusiveness of perfection. Being exposed to the normal bodies of others in an unclothed state, however, helps to combat this skewed body perception. When one is accustomed to seeing normal bodies, as Califia notes, a wider range of body types can begin to seem attractive.

The possibility of learning firsthand about new erotic techniques is an important aspect of sex and play parties for many attendees—indeed, for newcomers to BDSM, being trained by someone who knows how to use a flogger or a whip before using one on a human being increases the safety of such practices dramatically. Yet this practical element pales next to the psychological impact of having sex in front of others, and watching them have sex as well. Califia notes that play parties have made him more kind; I suspect that here he is referring to the effects of allowing one's awkwardness and vulnerability to be exposed in a safe environment. The experience of physical nakedness in company is secondary to the way it can represent (and sometimes cause) psychological nakedness: a person writhing in an orgasmic ecstasy, making the peculiar noises and silly facial expressions that characterize disinhibited sexual pleasure, is as unselfconscious and present-focused as a child. Sexual ecstasy has a kind of innocence because, in its throes, all ego is gone: all one's social polish, job title, or advanced degrees. With mussed hair or smeared makeup or boxers around ankles, sexual ecstasy provides an opportunity to truly let go. Public sex offers a place to have that vulnerable, naked self seen and accepted, not just by a trusted partner, but by those considerably less invested: friends, or even strangers. Such opportunities can be particularly challenging and transformative for men, who are often socialized to avoid showing vulnerability. People who have never before met can smile upon each other's flustered post-orgasmic languor with a sense of kinship: we are radiant in our pleasure, but sweaty and a little awkward; we are unapologetically fierce, or sentimentally tender, in ways that are not usually socially acceptable; we are all beautiful, and all flawed.

Most people in Western culture have never seen sex between people who love each other. In the stilted encounters that most mainstream pornographic films feature, it frequently seems that the actors may not even like each other, or at least that liking is irrelevant to their working relationship. But the affection, ferocity, playfulness, lust, or ecstasy that people can witness at sex and play parties are not media productions; they are real. There is the potential for awe in witnessing these interactions—the possibility of grace, and the potential for divine presence. Just as every lover we take can reveal something different about ourselves and how we are in relationship, there is a mystery in witnessing ecstasy, the potential to glimpse something unspeakably holy in a stranger's rapture.

In the 2006 film *Shortbus*, the main character—a sex therapist who is unable to have an orgasm—accepts an invitation to an underground salon where local alternative people gather for socializing, music, and sex. Standing clothed in the doorway to a room where perhaps a dozen couples and small groups are pleasuring each other, the therapist catches the eye of a woman who is ardently making love to her partner. The two lock gazes; the woman's eyes shine out of the frame at the audience, soft, open, yet intent. After a few moments, the therapist steps away, visibly moved and seemingly unnerved by the potent moment of contact. Although the interaction could be read as impersonal, as the therapist and the woman do not know each other, the therapist's reaction makes it clear that *something* has penetrated the shell of her isolation. Their encounter is deeply personal, yet not entangled with the context of any pre-existing relationship, agenda, or goal. Opportunities to experience others in such an unguarded and primal state are few and far between. Yet for someone in an open and seeking state, like the main character of *Shortbus*, sex and play parties that successfully create a safe space for desire can offer them up in abundance.

Within the healthiest of these erotic communities, sex and play parties encourage a culture of no-pressure flirtation and celebration. In the presence of clear rules around consent and community norms that sex is obtained only by asking directly, compliments and admiration are less likely to be misunderstood and so are given more freely. Participants' desirability and worthiness to receive pleasure are consistently affirmed. Such healing environments can send participants back out into their lives happier, more grounded, and less restricted by personal and sexual traumas. Sexual touch offered as a sacrament in a community setting can have a profound impact on those who have been wounded by negative social attitudes around sexuality and bodies. I am reminded of the Jewish notion of *tikkun*, or repair—the idea that the acts of human beings can heal the wounds of God. In a panentheistic Pagan model, when we restore our communities and our selves to health, we mend Being itself, for Being is reflected and made manifest in each one of us.

The Erotic in Capitalist Context

Although it is possible to create safe, respectful, and loving spaces for sexual expression and experimentation, sex radicals—as well as those who consider their sexuality to be more "garden variety"—are all impacted by the wider culture's attitudes toward sex. As discussed in Chapter 4, certain esoteric philosophies and psychoanalytic approaches have linked sexual and spiritual liberation. Today, elements of the idea that sexual freedom is linked with psychological or spiritual freedom have percolated throughout popular culture. Feminism may have been speaking from a politically leftist position when it first linked women's self-administered orgasms with empowerment and liberation from patriarchy, but similar trends have emerged even in some politically conservative communities. Many evangelical Christians are now touting "hotter sex" as a way for married couples to sustain their marriages and their faith,[16] and on a prominent Christian sex blog, wives advocate praying to Jesus for a "sexual awakening" so as to "walk in sexual freedom."[17]

Some critics of this movement within evangelical Christianity have panned it as a marketing scheme, however. About former megachurch preacher and married sex advocate Ted Haggard, one blogger notes, "Maybe he's thinking it's good evangelism: 'if you aren't satisfied with your sex life, come to church!' After all, sex is used to sell everything else, maybe he figured he could use it to sell Jesus."[18] The writer implies that hot sex does not actually deepen evangelical Christian faith, but has been incorporated into evangelical Christian teachings to keep from losing the faithful to Christianity's religious and secular competition. Similarly, Hugh Urban is also concerned about the commodification of sex in Western culture and its corrupting impact on those who seek sexual-spiritual liberation through more esoteric routes. He writes:

> [C]ritics like Herbert Marcuse have suggested that contemporary society is now based on a kind of manipulation and exploitation of sexuality in the interests of consumer capitalism. [. . . I]n the name of sexual "freedom" and the "liberation" of eros, we are now

> all in a sense enslaved to a consumer culture in which sex is used more to sell and make us buy things than to achieve any sort of real emancipation.[19]

Books, workshops, and tools for "tantric" practice are popular among those in the Pagan and New Age movements, and such products frequently promise spiritual development without necessarily advocating for the hard work of practical personal or political change. Urban's concern that one form of restriction (sexual repression) has simply been exchanged for another (addictive consumption of products that promise liberation) is worth considering.

Even those whose lives are devoted to sexual and spiritual exploration do not necessarily find satisfaction. For example, twentieth-century magician Aleister Crowley's body of spiritual writings focuses on the expression of desire and the defiance of social taboos. "The word of Sin is Restriction," declares Crowley's channeled scripture, *The Book of the Law*.[20] Yet Urban uses Crowley's lonely death to question the ultimate effectiveness of seeking liberation through extreme sexual or other physical practices. In Crowley's autobiographical novel, *Diary of a Drug Fiend*,

> King Lamus frees his disciplines by teaching them how to indulge in all manner of drugs, sex, and other vices, yet without becoming attached to them. That is, he allows them to experience all desires in order to free them of the tantalizing desire for their forbidden fruit. Yet while this strategy of freedom through indulgence works well in Crowley's novel, it does not appear to have worked so well in Crowley's own life. Rather, the Beast [Crowley] spent his last years largely in isolation, increasingly addicted to heroin and apparently exhausted of his own once infinite will to power. [. . .] Crowley and his characters epitomize [. . .] the recurring tendency for this ideal of sexual liberation to become mingled with less admirable sorts of things, such as misogyny, drug abuse, or simple commercialization.[21]

In contrast to the queer, polyamorous, and kink communities that Hutchins describes, Urban's sexual-spiritual pioneers are radical individualists. For Crowley and others, spiritual and sexual liberation derived from breaking taboos, claiming the right to pleasure, and achieving experiences of the transcendent. Notably lacking, however, was a commitment to empathy, mutuality, or community. Urban's criticisms of sexual liberation in the more community-oriented contemporary Pagan movement are accordingly milder. Urban frowns on the conventional gender roles and sexism that still plagued the Pagan movement in the twentieth century, but the objectifying attitudes that he finds among other pioneers of sexual magic do not appear in his analysis of Pagans.[22] The group solidarity felt by many contemporary groups of sexual minorities may help to control the destructive tendencies that Urban describes.

The misogyny, addiction, and commercialization that marred the careers of many sexual-spiritual pioneers in the nineteenth and twentieth centuries show that a direct assault on a culture's mores does not necessarily liberate one from their influence. The undesirable behaviors that Urban describes were all endemic to Western culture in the nineteenth and twentieth centuries. Although feminism has made some impressive shifts in women's political, social, and economic positions, Western culture is still far from gender-egalitarian. Addiction (to alcohol, drugs, food, gambling, sex, and other sources of excitement or pleasure) continues to warp many of our relationships, and high-pressure advertising for consumer goods is so ubiquitous as to be unavoidable. The negative impacts of these unhealthy social realities are experienced as normal, and we may notice them only when they reach an extreme level. When these addictive and commodifying tendencies manifest themselves in people who are actively trying to change existing social mores, however, their presence can become an ugly and obvious disfigurement on what otherwise might be a beautiful alternative vision.

Co-optation and Corruption of Sexual Minority Communities

I have spoken of the practices of healthy communities of sexual minorities, and of the Pagan movement as one spiritual umbrella under which body-positive religious people gather. Such communities are not without their risks, however. As Hutchins observes, newcomers face significant emotional challenges: they may be intimidated or overwhelmed by the opportunity for long-desired erotic freedom, leading to emotional shutdown or withdrawal; or they may take the freedom offered as permission to skip over normal, healthy processes of relationship-building, leading to an intrusive emotional pattern Hutchins calls "premature merging."[23] In the absence of sufficient community mentoring, newcomers may also carelessly violate community norms. Hutchins observes that when leadership is weak in erotic communities, sex and play parties may lose their group bonding aspect, leading to recreational sex that fails to build intimacy or support personal growth (which, while perhaps not an inherent problem, at least represents a significant missed opportunity).[24] Most problematically, however, some sexual predators imitate erotic communities' rhetoric of honesty and consent in order to manipulate and take advantage of the vulnerable.[25] A predator, for example, might present himself as committed to open, honest communication while concealing the existence of other, non-consenting partners; mislead a potential partner about his STD status; or use the community's gossip and reputation-checking network to malign and therefore isolate a partner he wants to control or keep to himself. Such manipulations are impossible to completely prevent, but a well-bonded community with stable, long-term members can create an environment where newcomers are vetted and socialized to the community's norms.

Sexual minorities are also vulnerable to being co-opted by the broader culture, which does not share their values. For example, BDSM practitioner Stacy Fowles worries that BDSM imagery and rhetoric are being actively abused by the mainstream pornography industry. Fowles describes how, during the pre-Internet flourishing of BDSM culture in the 1970s and 1980s, fetish pornography had to be painstakingly sought out by invested and informed viewers. Now that readily available Internet pornography contains images of bondage, flogging, or dominance/submission, Fowles fears that these decontextualized images train viewers to believe that women genuinely desire sexual coercion. In contrast, fetish pornography produced by the BDSM community actively deconstructs this belief by calling attention to the images' artifice. She writes:

> BDSM pornography is so excruciatingly aware of its own ability to perpetuate the idea that women yearn to be violated that it actually fights against that myth. At the end of almost every authentic BDSM photo set, you'll see a single appended photo of the participants, smiling and happy, assuring us that what we've seen is theater acted out by consenting adults, proving that fetish porn often exists as a careful, aware construct that constantly references itself as such.[26]

In some ways, Fowles' is the cry of every co-opted subculture: that the mainstream has stolen its counterculture aesthetic while paying lip service to (or even discarding) its associated values. Although BDSM education and advocacy groups have a growing Internet presence, casual consumers of pornography are unlikely to find these resources when doing a quick search for sexually explicit material. For Fowles and other feminist BDSM practitioners, the misuse of BDSM imagery is just one more reason that the kink community must speak publicly about its commitment to explicit consent. Without an understanding of BDSM's culture of consent and negotiation, she argues, the use of BDSM imagery in mainstream pornography could influence viewers to fetishize genuine violence.

In interviewing Pagans about sexuality in their communities, Jennifer Hunter also balances praise of the community's ethic of sexual openness with criticism. As one female Wiccan observes:

> In traditions with secrecy oaths, there is always the uneasy fact that some high priestesses or high priests use sex as a lever, in the manner that Starhawk would call "power-over": "You must have an actual sexual Great Rite [ritual intercourse] in order to move to the next initiation . . ." or "I will teach you if you have sex with me." As with any other religion, there are pitfalls. Paganism, having such an intertwined path between sex and religion, walks a careful balance that can be easily upturned.[27]

Not all those who abuse others in the name of religion or sexual liberation are consciously predatory. Those raised in dysfunctional or abusive families may engage in addictive or other destructive behaviors in an attempt to cope with past traumas. Further, otherwise well-intentioned people may not yet have recognized and challenged the subtly sexist, racist, or elitist attitudes that they absorbed from mainstream culture. These attitudes influence them to mistreat themselves and others in erotic contexts.[28] One polyamorous Pagan man told Hunter:

> Under the guise of "free love" and "sexual empowerment," a lot of people manipulate, deceive, dominate, and dump their lovers, then justify their actions with "I'm taking control of my sexual power." . . . I'm sure you've encountered such folks yourself . . . supposedly "enlightened practitioners of sacred sexuality" who are little more than addicts with a good pickup line. This sort of bullshit erodes the trust and openness that's essential for people to share a sacred sexual bond, and to maintain public credibility for these practices. (My roommate, no prude herself, has said, "These girlfriends of yours keep talking about how healthy and powerful they are, but their actions show just the opposite.")[29]

In an ideal world, an open heart and a commitment to ethical conduct would be sufficient preparation to safely challenge mainstream taboos about touch. Anecdotes like the above, however, demonstrate the psychological, emotional, social, and physical risks of such activities. Rubin remarks that "Sexual acts are burdened with an excess of significance,"[30] and perhaps this accounts for the strength of our society's taboos around sexuality in particular and touch in general. To break a taboo in the name of love and the sacred can be enormously freeing. Yet anxieties around the body and intimacy can cause the same act to release demons, both inner and outer. The girlfriends of the speaker above seem to be exhibiting this destructive pattern: although they seek to embrace the sacred through their non-mainstream sexual practices, their actions ultimately demonstrate their woundedness and their failure to achieve the connectedness and power that they seek.

Power and Vulnerability

To stay within the bounds of socially-acceptable behavior is often to minimize obvious harm. There may be fewer opportunities for ecstasy in a mainstream heterosexual marriage, but there are also fewer pitfalls and more hard-and-fast rules for what constitutes normal and ethical conduct. Once these rules have been discarded, the potential danger to self and others is greater (at least, until an alternative set of values and a community that supports them are solidly in place). Because sex is so heavily weighted with significance in our society, the psychological impact of its use and misuse can be profound. A first sexual experience can make a young person walk tall in its afterglow; a profound sexual rejection can destroy the same person's self-esteem and shadow his intimate relationships for years afterward; and a sexual assault can impair a person's functioning in all aspects of his life. Although Rubin suggests that too much meaning has been assigned to sex acts, perhaps the embodied nature of sex makes this "excess of meaning" inevitable—we cannot help but develop taboos, insecurities, and demons around something that can be so intimate, ecstatic, and powerful. Such power, for good or ill, is self-perpetuating. A positive experience of touch holds all the potential for joy and healing that a negative one holds for damage and harm. Contrary to many of the messages we receive from the media, it is rare that sex can ever be "*just sex*," no matter how casual we attempt to make it.

Indeed, sometimes the dual nature of power—the way that joy and agony are two sides of the same coin—is revealed when the pursuit of ecstasy is most successful. In *Radical Ecstasy*, Easton and Hardy describe their experiences of transcendence while engaging in extreme BDSM practices. But the authors confess that writing the book produced some unexpected, difficult side effects: an openness so profound that it became vulnerability, an uncomfortable *skinlessness* that impacted their daily lives. Hardy writes:

> For the last two years, we have both written about all of our SM play as part of the process of writing this book. That means, for the last two years, I have done no scene that has not been about

> pushing my own edges, and after which I have not then required myself to examine those edges under a microscope and note my findings in a lab notebook for the perusal of others. [. . .] For me, it seems as though everything inessential in my life is closing down, bit by bit, as I focus all I've got on this single task. I haven't been able to read a book for pleasure in many months. I am doing no casual play and very little masturbation. I have rearranged my life so that I spend most of my time in solitude—my only employee now works elsewhere, my son sleeps different hours than I do and my lover comes over only on weekends. My skinlessness makes it difficult for me to talk about anything but trivial matters with anyone but Dossie, and my conversations with Dossie are almost always tearful because I cannot talk about anything that is not deep.[31]

In traditional India it is possible to live as a holy person, utterly without possessions and dependent upon the generosity of a temple or a patron. Such renunciants can spend their days in a state of meditative *samadhi*, the blissful contemplation of the divine. Their basic needs are provided by those who wish to interact with someone who is engaged in the pure pursuit of the transcendent. Westerners, however, have no such role available; when Hardy and Easton pursued ecstasy through touch and desire into a state of skinlessness, the only protective monastery available was the one they could create themselves. Inevitably, both had to regrow their boundaries enough to return to their jobs and families, to again function in the wider world as writers and professionals. It is possible, in the pursuit of the sacrament of touch, to achieve such a sustained state of intensity that it becomes difficult or impossible to function in daily life. As a Pagan rather than a Hindu, however, I cannot become a renunciant; my embodiment is intended for a sacred purpose, and to permanently retreat from society, family, and friends would constitute a rejection of the holiness of those connections.

The sacrament of touch requires *boundaries*. If there is no difference between self and other—if we indeed become permanently skinless—contact becomes impossible. In Chapter 1, I quoted Crowley's *The Book of the Law*, where the Goddess says, "I am divided for love's sake, for the chance of union."[32] There is no transcendence without something to transcend, no union without difference. In performing the sacrament of touch, those of us who are most successful in freeing our egos and experiencing ecstatic union encounter one last, subtle danger—that of opening ourselves so completely to connection and desire that we lose our boundaries, lose the very selves with which we touch. There is a temptation in ecstasy to attempt to remain in a place of boundaryless bliss, but in so doing, we cease to be priestesses or priests: we can no longer help to bring others into a place of connection and empathy; we are no longer the hands of the divine on earth. Like Buddhist *bodhisattvas*, we must turn away from skinlessness and ecstatic dissolution in order to be present here and now, feet firmly touching the ground, swirling in the steps of the erotic dance that is Being.

As I mentioned earlier in this chapter, touching in a state of conscious embodiment also brings us face to face with our mortality. In its most intense forms, erotic ecstasy can allow us to encounter death in a slightly different form: as dissolution, as the end of the self. The French called the orgasm *la petite mort*, the "little death"—an apt way to describe how sexual climax allows us momentarily to forget ourselves. Califia writes of the way ecstasy and the possibility of death can sometimes be unhealthily intertwined—and indeed, perhaps risking one's life always breaks a powerful taboo and pushes the transgressor closer to a moment of transcendence. Califia, however, argues that the act of maintaining some boundaries does not cut one off from ecstasy. Rather, the ability to hold boundaries while seeking love and erotic freedom is a sign of maturity and an affirmation of the will of the divine. On the topic of learning to eroticize sexual responsibility and safer sex in the age of AIDS, he writes:

> I know that the desire to experience bliss or connect with another human being comes from a good place. I've stuck needles in my arm to get there. I've taken cum and blood into my body because I had to get out of that little box called reality or die of loneliness. I will never apologize for any of the risks I've taken. But if those of us who value queer bodies and pagan transcendence do not survive, who will explain the morality that we lived by? If some deity made me, s/he wanted me to have ecstasy, not self-immolation. [. . .]
>
> We've been taught to believe that we can only experience ecstasy by doing things that might kill us. The supposed choice between intense but destructive pleasure or a healthy and virtuous but extremely boring life is a false dichotomy. [. . .] It is a radical act to take responsibility for your own and another's well-being in the face of hatred and prejudice. Why don't we see the people who do this as sexy, daring outlaws? Why doesn't our community applaud and reward such secular saviors?[33]

Performing the sacrament of touch, whether through simple affection or through radical sexual practices, is a risky undertaking, one that has many dangers. Yet as Califia points out, practitioners need not fetishize needlessly risky behavior, nor assume that "real" ecstasy can only be had when one's life is tangibly on the line. Well-considered boundaries are a sign of health, not of weakness.

Babalon's Cup

I have spoken of the dangers of touch in this chapter, and if these have given my readers pause about the wisdom of engaging in some of its forms, this hesitation was part of my aim. In the end, however, I hope that some readers will take up the obligation of the sacrament: that as a priest or priestess of the erotic divine, one is called to risk in the name of joy. As a massage therapist, I have lain my hands on the naked bodies of strangers to soothe and heal; as a lover, I have risked disease for the glory of connection. Like Califia, I can honestly say I regret nothing. There is no greater gift than the smile on a client's face when she gets up from the table, her pain relieved; there is no greater pleasure than good sex, anything from simple affection to the kind of lovemaking that engages me body and soul, sex that tears the veil off the face of the universe or lets me see my partner in her unmitigated glory. Eros is one of the names of God Herself, and touch is the practice of my religion.

In Chapter 1, I wrote of the Goddess as the Great Whore, one who understands and accepts all beings carnally, intimately, and unconditionally, regardless of their worthiness or condition. And though this is a holy image to me, it is also a fearful one: if the Goddess is a whore, to drink from her cup can never be without risk. Human beings may be her creatures, but so too are her viruses, her bacteria, all the tiny beings of the natural world that are an essential part of our ecosystem, but whose presence can mean death to humans and animals. Once, while making a decision about my sexual health, I felt Her before me, offering me a cup of blood that was teeming with living things, life and death birthed from the same womb. And she asked me, *Are you my priestess? Is your love stronger than your fear of death?* I hesitated for a moment: I consider the risks I take, and I do not endanger my loved ones without their consent. I place limits on the expression of my love . . . but I do so out of love for myself.

My answer is yes. I drank from the cup.

Chapter Summary

Chapter 5 confronts the considerable risks that come from defying restrictive social norms around touch and what constitutes acceptable sexuality. Those who seek to be more sexually open than the norm can become the special targets of sexual predators; people who would be activists for erotic liberation should cultivate good practices of self-care and form alliances with like-minded others for support and protection.

Nonstandard erotic behavior also has emotional risks. Defying standard social norms around touch can destabilize existing relationships or bring people face to face with mortality in the form of diseased, aging, or injured bodies. Though there are real medical risks to both nonsexual and sexual touch, those seeking more or different touch in their lives may also deal with negative social judgments, such as the sensationalism and hysteria around even minor and non-life-threatening STDs. Social attitudes around such diseases sometimes function more as emotionally-charged purity taboos than informed and considered health concerns.

Alternative erotic communities can provide education, emotional support, and safe spaces for erotic expression for sexual minorities. As with any other alternative community, however, they must actively police themselves for predators and are sometimes threatened by co-optation and commodification by those with commercial interests. Further, newcomers to such communities often arrive with an incomplete understanding of the sexual ethics they wish to embrace, as well as emotional burdens of trauma and shame, all of which may lead them to hurt or endanger themselves or others before they are able to heal. Such sexual minority communities nevertheless provide spaces in which alternative erotic norms can be developed and tested. I end by recommending the embrace of calculated erotic risks.

Conclusion

UNCOVERING THE FLESH OF THE WORLD

> *So many things can go awry with our fragile and unpredictable bodies. And yet, the flesh is the primal fountain of all our enjoyments, even the most esoteric: muscle and bone, anus and mouth, nerves and blood, fingers and toes, organs and tendons, brain and skin, semen and sight.*
>
> Patrick Califia, *Speaking Sex to Power*, p. xxi

Traditional Western theology has often treated the physical as if it were secondary, subordinate to a hypothetical state of pure consciousness. Yet as far as I can discern, no consciousness exists without a relationship to the physical. The body is the vehicle of all our experience: pain and pleasure, thought and emotion, action and rest. Even those states we think of as "spiritual" remain tethered to our physical bodies; our travels in dream, imagination, or in the realms of spirit cannot be integrated into our lives until they are encoded in the living flesh of the nervous system and brain. As human beings, all perception, all relationship, all knowledge and experience of the divine comes through our bodies. Acknowledging that the physical is a reality and perhaps even a necessity, we may discover the body as a divine theophany, the fleshly matrix of God Herself through which consciousness constantly evolves. Yet few of us are skilled in giving our bodies the balanced care that we need for physical, mental, and spiritual health, and our efforts are compromised by social structures that denigrate and ignore bodily needs.

Much existing erotic theology has focused on human sexuality and sexual ethics, which are important areas of inquiry. Sexual urges naturally ebb and flow throughout a human life, however, and not every reader will find sexual liberation to be a personally compelling topic. Yet touch is potentially part of all of our daily lives, and the neglect of loving nonsexual touch is causing tremendous harm to our society physically, emotionally, and socially. By focusing on it here in the context of a larger erotic theology, I hope that the concreteness of touch will help readers approach the complex flow of the erotic in their own lives.

For reasons that are rooted in our biology but that extend to the most complex aspects of our consciousness, loving touch is necessary for human flourishing. This book presents a cosmology, a basic system of ethics, and the outline of a spiritual practice that places pleasurable touch at the center of human life. In an environment of honesty and mutuality, we can bring intentionality to everyday touch—whether handshakes in the workplace, hugs among family or friends, or bedroom play with lovers—such that ordinary touch becomes a sacrament, an outward and visible sign of divine presence. From conscious experiences of relationship, known through the fragile vehicle of the body, a new set of values can emerge and inform our decision-making on both personal and social levels. Pleasure and the autonomy of the individual body become basic human rights and must be considered priorities when we establish group policies and allocate resources. The notion of suffering is no longer limited to gross physical or emotional injury; neglect and deprivation come into focus as social sins. Further, if pleasure is considered a human right, the patterns of exploitation that arise from rigid social, political, and economic inequality become intolerable, as does discrimination on the basis of benign erotic variation. Rather than condemning all relationships with power differentials as exploitative, however (for a society of perfect equals is surely impossible), an ethic of sacramental touch provides a framework with which to approach consent, empathy, and the negotiation of power exchange. Honoring the autonomy and personal experience of individuals, we become obligated to respect the testimony of adults who declare themselves to be in loving relationships, regardless of our personal erotic tastes. Nor is this principle of respect merely for those whose tastes are conventional; those with more exotic sexual proclivities must also respect others' preference for gentle, heterosexual intercourse, or for no sexual activity at all, rather than treating it as a sign of unhealthy inhibitions.

Why a Pagan Erotic Theology Now?

Although my own professional and spiritual evolution provides the immediate context for this book, it also emerges out of the particularities of the American cultural moment. Debate over legalizing same-sex marriage continues to rage in the United States. I am privileged to live in Massachusetts, where same-sex marriages have been legally recognized since 2004, but as I write this in September 2012, I am delighted by this morning's news that the city council of my hometown of Austin, Texas voted unanimously to endorse their legalization.[1] As states continue to pass and then pull back laws recognizing same-sex marriages, pockets of support even in largely conservative states give me hope that legal recognition will be national within a generation, if not sooner. Yet the struggle for marriage rights is just the most visible part of a larger movement for the fair treatment of sexual minorities. In urban gay pride parades around the nation, bisexual and transgender advocacy groups have become much more visible than they were a decade ago. Continuing to curtail discrimination against sexual minorities, however, requires the spread of a pluralistic erotic ethics and a wider acceptance of consensual, loving relationships between adults. Exploring the ethical ramifications of treating pleasure as a human right is one way to contribute to the fair treatment of sexual minorities.

Additionally, this theology is being written at a special moment in contemporary Paganism. Paganism is in its adolescence as a religious movement, and its development is being strongly influenced by communications technologies. Even as Pagans who are geographically isolated have gained access to resources that were unavailable twenty years ago, that same easy access to information about Paganism and to other Pagans via the Internet has, in some cases, impeded the growth of stable in-person religious communities. Although there are strong communities of Pagans in a number of urban areas, in general the contemporary Pagan movement is scattered and somewhat diffuse, leading many Pagans to become frustrated with their lack of access to competent religious services (for example, support for rites of passage, pastoral counseling, devotional rituals led by trained clergy, spiritual and magical training, etc.). However, because of a historical commitment among Wiccans not to charge money for training in their religion, as well as a wider Pagan suspicion of organized religion, few Pagan communities are currently willing to support paid, professional clergy.

Within the Pagan movement, tension is growing between those who are pushing for institutionalization—professional training for clergy;[2] the establishment of schools, libraries, nonprofits, and temples;[3] formal roles for elders to increase accountability within communities,[4] etc.—and those who believe that institutionalization will strangle their highly individualistic and potentially subversive spirituality.[5] As mentioned in the Introduction, some Pagans are also suspicious of formal theologies in the belief that where there is written theology, dogma and rigid doctrine inevitably follow. The challenge that contemporary Pagans face is to create structures that will support communities and coherent theologies without fatally compromising Pagans' diverse values.

Among those in favor of more structure, there is a temptation to put new wine in old wineskins: to shoehorn a protean new religious movement into the familiar organizational structures associated with mainstream churches and seminaries. These often-hierarchical structures do not always encourage the egalitarianism, collaboration, and diversity that many Pagans value. When paid clergy enter the picture, hierarchical structures threaten to strip the volunteers who built Pagan traditions and communities of their decision-making power. At the same time, however, initial experiments with non-hierarchical structures over the past thirty years have come under fire: for example, several former elders in the Reclaiming tradition have recently attacked its consensus process as lacking safeguards against bullying and failing to ensure ethical accountability.[6]

In order to sustain their communities, Pagans will need a variety of creative organizational structures and formal theologies, not just one. The recognition that Paganism is a movement consisting of many traditions and groups, each of which has a somewhat different set of values and needs, will help Pagans to focus their efforts on their own traditions or local communities, rather than on the movement as a whole. Yet the process will still require flexibility and experimentation, as well as a commitment to diversity. With this book, I seek to contribute to the intellectual structures of contemporary Paganism, but with the understanding that Pagan theology is necessarily in process, fluid, and organic. I want to open conversations and debates, not to have the last word. I am also sensitive to Pagans' pragmatism, their conviction that the true test of a belief is in its practical application. This theology is a framework that is open to the experience of mystery, but it seeks to integrate experiences of the ineffable into concrete practices and shared community norms. It is a theology of the erotic, an abstract category, but it is also a theology of touch, of the situational and specific; it examines the nature of the gods and God Herself, but it also explores the human body in relationship.

The rise of twenty-first-century communications technology is also part of the context for this theology. Psychology of technology scholar Sherry Turkle's recent book *Alone Together* describes how, in a time of vastly increased communication, many Americans are nevertheless experiencing a loss of human intimacy. The amount of information many of us take in every day is staggering: we skim blogs and news websites in the way people used to read the morning paper or watch the news, but we also regularly communicate with others via e-mail, texting, instant messenger, and social media. For those who carry smartphones or other mobile devices, this interaction may be nearly constant, consisting of dozens or hundreds of messages a day that cease only when the user is asleep. As Turkle describes, communications technology can function like an addiction; to be separated from one's mobile device, cut off from the flow of communication, is almost like a physical pain. For others, the communication is more like an assault, inducing a state of anxiety and overload—and yet because of job requirements or other circumstances, it may be professionally hazardous to log off and become unavailable.[7]

Turkle writes that when technology users are unwilling or unable to put their devices aside, they cope by developing distancing behaviors that restrict the intimacy and immediacy of the communication. Today's adolescents rarely use their mobile devices as phones, preferring instead to exchange series of brief text messages. Unlike a verbal conversation, a text conversation can be put on pause at any time: because the person on the other end of the communication never knows if their conversation partner is temporarily ignoring them or has simply been interrupted, one can withdraw from a conversation without rudeness. These pauses allow the sender to deal with (or avoid) uncomfortable emotions, or to try to make the best impression possible by carefully crafting a spontaneous-seeming message.[8] As Turkle points out, however, this situation is creating a generation that lacks the skills to have intimate verbal conversations. One adolescent that Turkle interviewed acknowledged his discomfort with verbal communication and indicated that, although he knew he would need to develop the skill to become a successful adult, he didn't feel ready for the stress and anxiety associated with higher-stakes in-person conversations.[9] Other interviewees remarked that they either avoided phone conversations because of the focused attention required, or because their friends reserve verbal conversations only for high-priority communication, like life crises.[10] Some adolescent respondents were frustrated with the way communications technology inhibited intimacy in their family lives. Several complained that their parents would not put their devices away during meals, which made it difficult for family members to focus on one another; one teen reported that her mother rarely stops using her device long enough to look at or speak to her, even for a greeting when her mother picks her up from school.[11]

Turkle is also concerned about the way that technological interactions are being used as substitutes for human intimacy. In both the United States and Japan, responsive robots are employed in elder care facilities to help comfort and entertain residents. One popular Japanese robot is a soft, furry baby seal that makes complex movements and sounds in reaction to touch; realistic baby dolls are also available. Elders who are given these dolls to interact with report developing feelings of warmth and affection for the robots, as well as improvements in their overall emotional state.[12] Yet in one disturbing anecdote, Turkle tells of a visit she had with an elderly woman, her three-year-old granddaughter, and one of the robotic baby dolls. The elderly woman had been struggling to interact with her high-energy, demanding, and somewhat unpredictable granddaughter. When presented with the doll while her granddaughter was also present, the elderly woman became distracted by the doll's compellingly realistic yet predictable reactions. The doll was easy to care for and reacted adorably to the woman's ministrations much as a human baby would; the human three-year-old, however, was difficult to please and threatened to overwhelm her grandmother's fragile reserves of energy. Although Turkle observed the woman affectionately cooing over and cuddling the baby doll when it was present, when she interviewed her after the encounter was over, the elderly woman seemed uncomfortable. Perhaps realizing that she had become distracted from her granddaughter, she expressed disinterest in the doll and contempt for the idea of forming an attachment to it.[13] Turkle argues that robots provide compelling substitutes for human intimacy because they are so much more predictable than human beings. She fears that as artificial intelligence continues to improve, the temptation to avoid demanding human relationships will become ever stronger. Further, she is concerned by the trend in both the United States and Japan to supply the elderly with comforting toys rather than paying caregivers to provide loving human touch (or, for that matter, caring for the elderly in a family setting).[14]

In a highly technological, high-stress environment, there are constant temptations to limit opportunities for human intimacy in favor of more predictable, technologically mediated interactions. Yet as we have seen from touch deprivation studies, human beings need frequent loving touch in order to be healthy. Professional bodyworkers can provide safe environments to receive touch. However, weekly massages are an expense that many cannot afford, and professional massages do not fully substitute for a habitual pattern of affectionate touch within a loving family or among a circle of close friends. Today, it is apparent that we are subjecting our bodies to demands that they were never evolved to bear. Stress-related illnesses, anxiety disorders, and depression pervade American society, and the state of overcommitment and near-overload in which many Americans live can make them averse to dealing with new emotional challenges. In order to give the erotic a central place in our lives, we must set conscious boundaries around our use of technology, boundaries that reflect a commitment to loving, intimate human relationships.

Among those in my spiritual community, the idea of the "media fast" has become increasingly popular as a way to focus on family and loved ones, care for one's body, or enjoy the outdoors. On a monthly, weekly, or daily basis, those of us who spend much of our work life sitting in front of glowing screens make a commitment to turn them off—perhaps for one hour a day, one day a week, or one weekend a month. Some set rules for mealtimes and social gatherings: families make commitments to silence their electronic devices or leave them at the door. Such mindfulness practices can restore a sense of control over our use of communications technology and establish a container for the sacred work of forming embodied relationships.

The Erotic in Ecological Context

Because I have focused on touch between human beings as a sacramental practice, I have found myself concentrating on the cosmological, the social, and the personal while neglecting the ecological. Yet ecotheology provides an important additional context for how neglecting the relationship between our bodies and the environment potentially stunts our humanity. Much of this neglect is related to a cultural shift toward indoor activities. Americans have been concerned about the use of television as a babysitter for decades, but television has been at least partially replaced by compellingly interactive video games and internet technologies. Today, access to these entertainments leads many children and adolescents to spend the majority of their time indoors, sheltered from the unpredictability of both the elements and human society. Particularly in urban environments, parents keep children indoors and constantly supervised, either because of safety concerns or because of community standards: several news stories in the past few years have documented incidents where parents allowed children to play outside or travel alone in what they believed was a safe enough urban environment, only to have other parents report them to the police or to Child Protective Services for child endangerment.[15] In some regions, cultural factors also discourage spending time outdoors. A friend who works as an eighth-grade teacher in a low-income Massachusetts school district reported to me that his students use the word "country" as an insult (meaning rural, backwards, and unfashionable), and that they consider the practice of going camping to be little short of insane. For these students, the idea that one can form a positive relationship with the environment is nonsensical. Yet even among those Americans who value outdoor activities, there is a problematic separation between human beings and their environment: being outside is often framed as "getting out into nature," a process that requires traveling to a less-inhabited area, not cultivating a relationship with the land just beyond one's front door.

In *The Spell of the Sensuous*, David Abram suggests that this disconnection from the land and from the plants and animals that surround us may be contributing to an underdevelopment of human consciousness. Abram's discussion is framed by Western writing on phenomenology and language, but its urgency comes from his experience doing anthropological field work in Bali with traditional shamanic practitioners. During his year in Bali, Abram found his perceptions gradually becoming open to what he calls "the sensuous world": a world where all living bodies are in active relationship and communication with each other. Linking "the rapid deterioration of wild nature [and] the steady vanishing of other species" with "the flattening of human relationships,"[16] Abram envisions the body as inherently relational:

> The eyes, the skin, the tongue, ears, and nostrils—all are gates where our body receives the nourishment of otherness. This landscape of shadowed voices, these feathered bodies and antlers and tumbling streams—these breathing shapes are our family, the beings with whom we are engaged, with whom we struggle and suffer and celebrate. For the largest part of our species' existence, humans have negotiated relationship with every aspect of the sensuous surroundings, exchanging possibilities with every flapping form, with each textured surface and shivering entity that we happened to focus upon. All could speak, articulating in gesture and whistle and sigh a shifting web of meanings that we felt on our skin or inhaled through our nostrils or focused with our listening ears, and to which we replied—whether with sounds, or through movements, or minute shifts of mood. [. . . W]e are human only in contact, and conviviality, with what is not human.[17]

The notion of reciprocity with the non-human echoes language used by scholars of indigenous animisms such as Irving Hallowell and Graham Harvey, who speak of reciprocal relationships with "other-than-human persons"—and indeed, Abram's descriptions of the living world reflect many kinds of nonhuman consciousness. For Abram, humanity fully realized depends on relationship with this living environment. Without it, we lack access to experiences of depth and ecstasy that should be our birthright as living beings, and we also endanger ourselves and our descendants through wanton exploitation and destruction of the ecosystem.

Abram does not argue that Western civilization needs to do away with its technologies, but he does observe that awareness of the communicating, sensuous world is difficult to maintain in a Western urban environment. In fact, he relates that his own sense of connection to the wild started to slip within weeks of returning to the United States from Bali.[18] Technology has given us a greatly increased ability to control our surroundings, and combined with changes in perception associated with a literate rather than oral civilization, we find ourselves alienated from the expansive and intimate consciousness that Abram believes our ancestors enjoyed. By tracing some of the habits of thought and behavior that isolate human beings from their environments, Abram explores possibilities for human beings to reclaim the deeply connected style of perception he describes—but without completely giving up the advantages gained through Western technology and culture.

Reclamation of a connected consciousness begins and ends with the body and, more generally, with the concrete physical realities in which we live. Abram writes that at the end of his career, phenomenologist Edmund Husserl sought to ground philosophy and science in direct human experience rather than on transcendent hypotheticals.[19] The identification of the perceiving self with the biological body, rather than with an incorporeal intellect or soul, is a result of this shift in emphasis.[20] Abram then draws on the work of Maurice Merleau-Ponty to explore the consequences of this move. For Merleau-Ponty, the body is inherently part of a web of relationships with other bodies, simultaneously perceiving and perceived; consciousness arises from participating in this exchange.[21] As Abram explains, the nature of direct perception is necessarily reciprocal:

> [A] wholly immaterial mind could neither see things nor touch things—indeed, could not experience anything at all. *We* can experience things—can touch, hear, and taste things—only because, as bodies, we are ourselves included in the sensible field, and have our own textures, sounds, and tastes. [. . .T]o touch is also to feel oneself being touched [. . .] to see is also to feel oneself seen.[22]

Where I speak of the erotic, the intimate flow of life force that passes from body to body, Abram speaks instead of communication between perceiving bodies. For Abram, language is itself a bodily phenomenon, "a sensuous, bodily activity born of carnal reciprocity and participation." Here, touch is but one form of reciprocal sensory contact that is constantly taking place among an interconnected network of beings. Merleau-Ponty writes collectively of these bodies as "the Flesh," encompassing both human flesh and "the flesh of the world."[23] This terminology emphasizes the integrated nature of perception—the notion that all bodies are part of an intimate web, constantly influencing each other and making adjustments both minute and dramatic in response. By flattening this environment with technological controls and voluntarily isolating ourselves from communication with the nonhuman, argues Abram, we are in essence dulling our ability to perceive. Like the baby mammals whose physical development is stunted by lack of loving touch, our ability to perceive the nonhuman world—and ourselves—has become radically constricted by lack of contact with the land.

By foregrounding the body as a unique and necessary vehicle of experience, erotic theology supports us in reclaiming a sensuous relationship with the environment. Through making room for the experience of intimacy—with each other, but also with animals and plants, insects and birds, wind and water—we bring intentionality to our perceptions and potentially sacralize the physicality of the human condition. We need the conscious experience of being embodied, of touching and being touched, to make us fully human; and as Abram suggests, to be fully human is to be aware of one's engagement in the interdependent sensuous world, to open oneself to the perception that the human is in every moment shaped and reshaped by countless relationships. We are changed by our human interactions, but also by the food we eat, made from the bodies of animals, or grown from plants pollinated by dwindling hives of honeybees; our mental and physical health is impacted daily by molds and pollen in the air, as well as by pollutants, by the number of hours we spend in sunlight, by humidity and rain and the wind that brings it. This web of interactions is what Merleau-Ponty thought of as the Flesh, and it is also what many pantheist or panentheist Pagans think of as the Goddess or Gaia, the complex organic system that sustains all life on this earth. Interdependence is no metaphor, but a concrete reality that acts daily on our physical being.

Sacramental Touch and Body Mysteries

Abram believes, and I agree, that the ability to have intimate, erotic interactions with the environment and the beings we encounter is potentially healing for us as individuals and as a society. But the fact remains that Abram's personal strategy for achieving this consciousness—spending a year studying shamanism in Bali—is inaccessible to most Westerners, and perhaps even ultimately undesirable. The perceived need to access special spiritual knowledge "out there" in some exotic culture rather than in our immediate surroundings can inhibit the development of intimacy with the local. Still, for many of us, the methods required to achieve and sustain an erotic consciousness are obscure, and the experience itself remains elusive. It is tempting to suggest that, if achieving such perception requires extraordinary means, then it cannot truly be foundational to being human. Yet perhaps neither the means nor the experience itself need be extraordinary; perhaps they appear so only from a point of view that does not automatically identify the self with the body. A practice of sacramental touch could help us understand the self and the body to be one, and so provide a means to discover the erotic.

Many contemporary Pagans believe that one purpose of religious practice is to reveal hidden or occult knowledge (about divinity, or about divine/human relationships). As used in the Western esoteric tradition, the term "occult" has tended to refer to elite knowledge accessible only after a process of initiation. In contemporary American society, however, the term has been rendered nearly useless through its popular association with (largely fictional) Satanic cults. The practice of seeking hidden knowledge through unusual modes of consciousness continues to be important in contemporary Paganism, however. Theologians such as Constance Wise have worked to define "occult" in a way that better reflects contemporary Pagan and feminist values. Wise defines occult knowledge as the creative, non-rational, subliminal knowledge that arises from the experience of the human body in relationship.[24] Although she speaks specifically in terms of women's bodies, her concept is easily applicable to people of all genders. For Wise, occult knowledge is obtained through experiential means, including ritual practices and intuitive techniques such as divination or artistic production. Through these, practitioners encounter "mysteries," defined by Wise as "experiences whose meaning is essentially non-rational and cannot by conveyed by description and analysis alone."[25] The knowledge gained by encountering mysteries, then, is inherent to having a human body (or perhaps more broadly, having any kind of body), yet has been suppressed and made "subliminal" under oppressive social structures. Yet Wise does not necessarily argue that this knowledge takes the same form for all human beings. She is quick to point out that embodied knowledge is "relational, communal, and particular to a given social and historical context [. . .] shaped by factors such as race, class, sexual orientation, age, and group history."[26] In order for an experience of mystery to become knowledge, it must be situated in time and place, known in a community of bodies. When Abram speaks of encountering the very literal interdependence of the sensuous world, he is speaking of occult knowledge in Wise's sense—it is a knowing that cannot be learned from a book, and it goes beyond intellectual belief into a certainty that comes from the flesh. Yet it is knowledge that could utterly change how we relate to the land on which we live, just as conscious experiences of erotic flow could forever alter our relationships, spirituality, and politics.

I have left my vision of sacramental touch deliberately loose, as such a practice is necessarily personal and must be adapted to individual and community circumstances. Seeking the erotic love of the divine led me to become a professional bodyworker, but for others, the changes may be far more subtle. As a practical way of championing human beings' right to pleasure and bodily autonomy, however, bringing awareness and intentionality to touch is a practice that can integrated into any profession—and yet to constantly maintain that awareness, to consider issues of consent, and to treat others' bodies with respect can be a constant struggle in a culture where bodies are routinely violated in minor ways. To bring consistent presence to human touch is difficult enough; to also regularly be present to interaction with the natural world is still more complex. It requires a non-rational approach perhaps because the task is simply too complicated for the conscious mind. Yet this communication and connection is something our animal bodies are adapted for, if we can expand our perceptions.

I end here with a call to bring a practice of sacramental touch, not just into our families, but into our wider communities: to nurture both self-knowledge and desire, so that we can confidently say "yes" and "no" to touch; to cultivate discussions about consent, boundaries, and autonomy; to develop empathy and model mutuality; and to affirm that we are divine, yet made human for love's sake—born for pleasure.

Notes

Introduction: Pagan Erotic Theology

1. Gil Hedley, *Reconceiving My Body* (Xlibris, Self pub., 2000) 20–21.
2. Francesca De Grandis, *Be A Goddess!* (San Francisco, CA: HarperSanFrancisco, 1998) 33.
3. James Gleick, *Chaos: Making a New Science* (New York: Viking, 1987).
4. Marvin J. Ellison, *Erotic Justice: A Liberating Ethic of Sexuality* (Louisville, KY: Westminster John Knox Press, 1996) 118–119.
5. Jason Pitzl-Waters, "Parsing the Pew Numbers," *The Wild Hunt*, February 26, 2008, http://wildhunt.org/blog/2008/02/parsing-pew-numbers.html, referring to Pew Forum on Religion and Public Life, Church Statistics and Religious Affiliations, 2008, http://religions.pewforum.org/affiliations.
6. This neologism was originally introduced by twentieth-century ceremonial magician Aleister Crowley.
7. This list of characteristics appears in somewhat different form in Christine Hoff Kraemer, "Introduction: Pagan Theologies," *Seeking the Mystery: An Introduction to Pagan Theologies* (Englewood, CO: Patheos Press, 2012) xi–xiii.
8. See my discussion in Chapter 4 under the heading "A Brief History of Sacramental Bodywork and Healing Touch" (119–122) and Chas S. Clifton, *Her Hidden Children* (Lanham, MD: AltaMira Press, 2006) 58–66.
9. Doreen Valiente, *Charge of the Goddess: The Mother of Modern Witchcraft* (Brighton, UK: Hexagon Hoopix, 2000) 55.
10. For an overview of the development of attitudes toward gender and sexuality in contemporary Paganism, see Christine Hoff Kraemer, "Gender and Sexuality in Contemporary Paganism," *Religion Compass* 6: 8 (2012): 390–401.
11. For a more thorough treatment of the topic of Pagan theologies, see Christine Hoff Kraemer, *Seeking the Mystery: An Introduction to Pagan Theologies* (Englewood, CO: Patheos Press, 2012).
12. See Carol P. Christ, "Why Women Need the Goddess," *Womanspirit Rising: A Feminist Reader on Religion*, Eds. Carol P. Christ and Judith Plaskow (San Francisco, CA: Harper & Row, 1979) 273–287 (reprint); Starhawk, *The Spiral Dance*, 1979, 20th Anniversary Edition (San Francisco, CA: Harper San Francisco, 1999); Carol P. Christ, *She Who Changes: Re-imagining the Divine*

in the World (New York: Palgrave, 2004); and Constance Wise, *Hidden Circles in the Web: Feminist Wicca, Occult Knowledge, and Process Thought* (Lanham, MD: Altamira Press, 2008).

13. Charles Hartshorne's best-known book *Omnipotence and Other Theological Mistakes* (Albany, NY: State University of New York Press, 1984) is recommended as a good starting place for reading in process theology.
14. Michael York, *Pagan Theology: Paganism as a World Religion* (New York: New York University Press, 2003) 13.
15. York's *Pagan Theology* is a primarily taxonomic work that argues for the existence of a "pagan" worldview in broad terms. It does not attempt to present a systematic Pagan theology or ethics. York also does not explicitly take a practitioner's perspective when writing about Pagans and indigenous peoples—a significant difference from the other "theological" texts I engage here. Despite its title, *Pagan Theology* is best considered as a work of religious studies, not as a contribution to the discipline of theology (understood as a necessarily insider discourse).

    Additionally, although *Pagan Theology* represents an important contribution to Pagan studies, like Michael Strmiska and others, I am concerned with the way its choice of the term "pagan" conflates primarily white, urban, middle-class contemporary Pagans with economically disadvantaged, politically beleaguered indigenous peoples around the world, particularly when "indigenous" has been the self-identifier of choice for many of these groups when advocating for their rights (for example, see "2009 Parliament Statement of Indigenous People," *Council for a Parliament of the World's Religions*, December 9, 2009, available at www.parliamentofreligions.org/news/index.php/2010/07/new-release-pwr-statement-of-indigenous-peoples/). (See also Michael F. Strmiska, "Modern Paganism in World Cultures: Comparative Perspectives," *Modern Paganism in World Cultures: Comparative Perspectives* (Santa Barbara, CA: ABC-CLIO, 2005) 11–13.)
16. Jordan D. Paper, *The Deities Are Many: A Polytheistic Theology* (Albany, NY: State University of New York Press, 2005).
17. Erotic theology is related to feminist and queer spirituality movements, which tend to share values, political commitments, and some beliefs despite the fact that adherents come from multiple religious traditions. For example, on the similarities shared between queer spirituality movements in progressive Christianity and contemporary Paganism, see Yvonne Aburrow, "Is It Meaningful to Speak of 'Queer Spirituality?' An Examination of Queer and LGBT Imagery and Themes in Contemporary Paganism and Christianity," *Contemporary Christianity and LGBT Sexualities*, Ed. Stephen Hunt (Burlington, VT: Ashgate, 2009) 139–156.
18. BDSM is an acronym for "bondage and discipline/dominance and submission/sado-masochism." For a full definition, see p. 26.
19. Gayle Rubin, "Thinking Sex: Notes for a Radical Theory of the Politics of Sexuality," 1984, *Deviations: A Gayle Rubin Reader* (Durham, NC: Duke University Press, 2011) 181.
20. See Christine Hoff Kraemer, *Seeking the Mystery: An Introduction to Pagan Theologies* (Englewood, CO: Patheos Press, 2012).
21. See Judith Butler, *Gender Trouble* (New York: Routledge, 1990) and Michel Foucault, *The History of Sexuality: An Introduction* Vol. I, trans. Robert Hurley, 1978 (New York: Vintage Books, 1990).
22. Robert C. Fuller, *Spirituality in the Flesh: Bodily Sources of Religious Experience* (New York: Oxford University Press) 20–24.
23. Ellison 8.
24. Virginia Burrus and Catherine Keller (Eds.), *Toward a Theology of Eros: Transforming Passion at the Limits of Discipline* (New York: Fordham University Press, 2007) xiv–xv.
25. See, for instance, Michele Tracy Berger and Kathleen Guidroz (Eds.), *The Intersectional Approach: Transforming the Academy through Race, Class, and Gender* (Chapel Hill, NC: University of North Carolina Press, 2009).

26. Ann-Janine Morey, *Religion and Sexuality in American Literature* (Cambridge: Cambridge University Press, 1992).
27. See Tracy Clark-Flory, "Will Marriage Change Gay Love?", Salon.com June 28, 2011, www.salon.com/2011/06/28/gay_marriage_23/.
28. Christine Hoff Kraemer, *The Erotic Fringe: Sexual Minorities and Religion in Contemporary American Literature and Film* (Ph.D. Boston University, 2008) 252, citing Craig Thompson, *Blankets* (Marietta, GA: Top Shelf Productions, 2003) 53. See also Steve Jungkeit "Tell-Tale Visions: The Erotic Theology of Craig Thompson's *Blankets*," *Graven Images: Religion in Comic Books and Graphic Novels*, Eds. A. David Lewis and Christine Hoff Kraemer (New York: Continuum International Publishing Group, 2010) 337.
29. See Melissa M. Wilcox, "Innovation in Exile: Religion and Spirituality in Lesbian, Gay, Bisexual, and Transgender Communities," *Sexuality and the World's Religions*, Eds. David W. Machacek and Melissa M. Wilcox (Santa Barbara, CA: ABC-CLIO, 2003) 323–358.
30. Helen A. Berger, Evan A. Leach, and Leigh S. Shaffer, *Voices from the Pagan Census: A National Survey of Witches and Neo-Pagans in the United States* (Columbia, SC: University of South Carolina, 2003) 93.
31. Ramon Johnson, "Interview with Gary Gates of The Williams Institute," http://gaylife.about.com/od/index/a/garygates.htm.
32. For an accessible introduction to transgender issues, see Kate Bornstein, *Gender Outlaw: On Men, Women, and the Rest of Us* (New York: Routledge: 1994) and her more recent edited collection with S. Bear Bergman, *Gender Outlaws: The Next Generation* (Berkeley, CA: Seal Press, 2010).
33. As Yvonne Aburrow points out, however, androgyny plays an important role in British Wiccan theology, though this aspect is often neglected in the everyday understanding of Wiccan practitioners. See Aburrow, 147.
34. For more on gender and transgender in contemporary Paganism, see Christine Hoff Kraemer, "Gender and Sexuality in Contemporary Paganism," *Religion Compass* 6: 8 (2012): 390–401.
35. Raven Kaldera, *Hermaphrodeities*, Second Edition (Hubbardston, MA: Asphodel Press, 2009); P. Sufenas Virius Lupus, *All-Soul, All-Body, All-Love, All-Power: A Transmythology* (Anacortes, WA: The Red Lotus Library, 2012).
36. See Lawrence Foster, *Religion and Sexuality: The Shakers, the Mormons, and the Oneida Community* (Champaign, IL: University of Illinois Press, 1984).
37. See, for instance, Sierra Black, "My Successful Open Marriage," Salon.com, January 21, 2012, www.salon.com/2012/01/21/our_successful_open_marriage/.
38. For more on the relationship between Heinlein's novel and the Pagan practice of polyamory, see Christine Hoff Kraemer, "Contemporary Paganism, Utopian Reading Communities, and Sacred Nonmonogamy: The Religious Impact of Heinlein and Starhawk's Fiction," *The Pomegranate: The International Journal of Pagan Studies* 13: 1 (2011): 52–76.
39. Morning Glory Zell, "A Bouquet of Lovers: Strategies for Responsible Open Relationships," *Green Egg* #89 (Beltane 1990), reprinted 2006 on CAWeb: The Official Website of The Church of All Worlds, March 27, 2007, http://caw.org/content/?q=bouquet.
40. "Alt.Polyamory Frequently Asked Questions," Ed. Elise Mattheson, Internet FAQ Archives, September 9, 1997, March 27, 2007, www.faqs.org/faqs/polyamory/faq/.
41. Annalee Newitz, "Love Unlimited: The Polyamorists," *New Scientist* 191: 2559 (July 8, 2006): 44–47. Janet W. Hardy initially published under the pseudonym of Catherine A. Liszt.
42. Dossie Easton and Janet W. Hardy, *The Ethical Slut: A Practical Guide to Polyamory, Open Relationships & Other Adventures*, Second Edition (Berkeley, CA: Celestial Arts, 2009).
43. Berger 92, 144.

44. Jennifer Hunter, *Rites of Pleasure: Sexuality in Wicca and NeoPaganism* (New York: Citadel Press, 2004) 64–82.
45. Valiente 54–55.
46. See Raven Kaldera, *Pagan Polyamory: Becoming a Tribe of Hearts* (Woodbury, MN: Llewellyn Publications, 2005).
47. Hunter 147.
48. Hunter 147–148. Citing June M. Reinisch with Ruth Beasley, *The Kinsey Institute New Report on Sex* (New York: St. Martin's Press, 1990) 162–163.
49. Dossie Easton and Janet W. Hardy, *Radical Ecstasy* (Oakland, CA: Greenery Press, 2004).
50. Raven Kaldera, *Dark Moon Rising: Pagan BDSM and the Ordeal Path* (Hubbardston, MA: Asphodel Press, 2006) 16.
51. Robert C. Fuller, *Spirituality in the Flesh: Bodily Sources of Religious Experience* (New York: Oxford University Press, 2008) 135, referring to Ariel Glucklich, *Sacred Pain: Hurting the Body for the Sake of the Soul* (New York: Oxford University Press, 2001).
52. Patrick Califia, *Speaking Sex to Power: The Politics of Queer Sex* (San Francisco, CA: Cleis Press, 2002) xxi.
53. See, for example, Patrick Califia, "The Sadomasochistic Challenge to the Gay Marriage Paradigm," presented at the American Academy of Religion conference in Montréal, November 19–22, 2009. The text of Califia's abstract states, "Many Christian theologians and secular gay activists assume that legalized, monogamous pair bonds will cleanse same-sex relationships of the taint of sin or mental illness, provided the only form of erotic gratification for these spouses is vanilla (non-sadomasochistic [S/M or BDSM]) sex. Gay marriage is seen as an indispensable part of obtaining equal rights for lesbians and gay men. Monogamy is touted as the only way to redeem same-sex activity, enforce commitment to or responsibility for one another, and convince the larger heterosexual community that homosexuals can love each other. This discourse constructs sadomasochists as the Other and the Shadow—bad for public relations and spiritually bankrupt. S/M sexuality is decried as a form of harm to oneself and others, and identified with nonmonogamy. This paper will examine the possibility that nonmonogamy and BDSM sexuality constitute new sources of intimacy and committed relationships as well as spiritual growth."
54. Pat Califia and Drew Campbell (Eds.), *Bitch Goddess: The Spiritual Path of the Dominant Woman* (Eugene, OR: Greenery Press, 1998); Lee Harrington, *Sacred Kink: The Eightfold Paths of BDSM and Beyond* (Lynnwood, WA: Mystic Productions, 2009).
55. Mark Thompson (Ed.), *Leatherfolk: Radical Sex, People, Politics, and Practice* (Los Angeles, CA: Daedalus Publishing, 2004).
56. Dave Grossman, *On Killing: The Psychological Cost of Learning to Kill in War and Society* (New York: Back Bay Books, 1996) 12–15.
57. Judith Herman, *Trauma and Recovery* (New York: Basic Books, 1997) 54.
58. Herman 82–83.
59. Margaret Cho, "Forward," In *Yes Means Yes! Visions of Female Sexual Power and a World without Rape*, Eds. Jaclyn Friedman and Jessica Valenti (Berkeley, CA: Seal Press, 2008) 3.
60. Hazel/Cedar Troost, "Reclaiming Touch: Rape Culture, Explicit Verbal Consent, and Body Sovereignty," In *Yes Means Yes! Visions of Female Sexual Power and a World without Rape*, Eds. Jaclyn Friedman and Jessica Valenti (Berkeley, CA: Seal Press, 2008) 171–178.
61. Morten Ebbe Juul Nielsen, "Safe, Sane, and Consensual—Consent and the Ethics of BDSM," *The International Journal of Applied Philosophy* 24: 2 (2010, 265–288): 272.
62. Nielsen 272.
63. Nielsen 273.
64. Nielsen 273.

65. Nielsen 273.
66. Nielsen 275–280.
67. Nielsen 281–282.
68. Nielsen 279.
69. Nielsen 278.
70. Nielsen 279.
71. Nielsen 285. Committed BDSM relationships where top and bottom roles are maintained on an ongoing basis might be challenged as ultimately undermining rather than increasing individual autonomy. Such "24/7" relationships are relatively rare, however, and should be considered as a special case. See Chapter 2, Endnote 37.
72. Gabriele Hoff and Richard A. Sprott, "Therapy Experiences of Clients with BDSM Sexualities: Listening to a Stigmatized Sexuality," *Electronic Journal of Human Sexuality* 12 (September 20, 2009), www.ejhs.org/Volume12/bdsm.htm.
73. American Psychiatric Association, *Diagnostic and Statistical Manual of Mental Disorders*, Fourth Edition, Text Revision (Washington, DC: American Psychiatric Association, 2000) 572–574; Richard B. Krueger, "The DSM Diagnostic Criteria for Sexual Masochism," Archives of Sexual Behavior, American Psychological Association, 2010, www.dsm5.org/Research/Documents/Krueger_ASB%20Feb%202011.pdf.
74. Juliet Richters et al., "Demographic and Psychosocial Features of Participants in Bondage and Discipline, 'Sadomasochism' or Dominance and Submission (BDSM): Data from a National Survey," *Journal of Sexual Medicine* 5 (2008): 1660–1668.
75. Juliet Richters, quoted in Australian Associated Press, "Kinky? You Can't Beat It," *The Sydney Morning Herald* April 16, 2007, www.smh.com.au/news/national/kinky-you-cant-beat-it/2007/04/16/1176696736407.html.
76. Stacey May Fowles, "The Fantasy of Acceptable 'Non-Consent': Why the Female Sexual Submissive Scares Us (and Why She Shouldn't)." In *Yes Means Yes! Visions of Female Sexual Power and a World without Rape*, Eds. Jaclyn Friedman and Jessica Valenti (Berkeley, CA: Seal Press, 2008) 120.
77. Catharine MacKinnon, "Are Women Human?" Interview with Stuart Jeffries, *The Guardian* April 11, 2006, www.guardian.co.uk/world/2006/apr/12/gender.politicsphilosophyandsociety.
78. Andrea Dworkin's *Intercourse* (New York: Basic Books, 1987) has often been read as making this argument due to statements such as "Intercourse is the pure, sterile, formal expression of men's contempt for women" (175). Dworkin has denied this interpretation, however, seemingly on the basis that such statements describe the way intercourse has been culturally constructed, not the way it could or should be. The statement that "all sex is rape" has also been misattributed to Catharine MacKinnon.
79. Hanny Lightfoot-Klein, "The Sexual Experience and Marital Adjustment of Genitally Circumcised and Infibulated Females in The Sudan," *The Journal of Sex Research* 26: 3 (August 1989): 375–392, www.fgmnetwork.org/articles/sexualexperience.php; Comfort Momo, "Female Genital Mutilation," *Female Genital Mutilation*, Ed. Comfort Momo (Abingdon, UK: Radcliffe, 2005) 5–12.

1. Divided for Love's Sake: An Erotic Cosmology

1. Ben E. Benjamin and Cherie Sohnen-Moe, *The Ethics of Touch* (Tucson, AZ: SMA, 2005) 105.
2. C.G. Jung, *Collected Works* Vol. 10, *Civilization in Transition*, Second Edition (New York: Bollingen, 1970) 254.

3. Audre Lorde, "Uses of the Erotic: The Erotic as Power," *Sister Outsider: Essays and Speeches by Audre Lorde* (Freedom: Crossing Press, 1984) 57, 58.
4. Carter Heyward, *Touching Our Strength: The Erotic as Power and the Love of God* (San Francisco, CA: Harper & Row, 1989) 25.
5. T. Thorn Coyle, *Evolutionary Witchcraft* (New York: Penguin, 2004) 120.
6. Miriam deCosta-Willis, "Introduction." In *Erotique Noire: Black Erotics*, Eds. Miriam deCosta-Willis, Reginald Martin, and Roseann P. Bell (New York: Anchor Books, 1992) xxix. Cited in Robert C. Fuller, *Spirituality in the Flesh: Bodily Sources of Religious Experience* (New York: Oxford University Press, 2008) 99.
7. Starhawk, *The Spiral Dance: A Rebirth of the Ancient Religion of the Great Goddess*, 20th Anniversary Edition (1979; San Francisco, CA: HarperSanFrancisco, 1999) 14.
8. Heyward 25.
9. Starhawk, *Spiral Dance* 41.
10. I suspect that Starhawk understood this myth in monist and soft polytheist terms: the Goddess is the fundamental substance of the universe, of which all other deities and all other beings are ultimately part. I tend to read the myth instead through the Western esoteric tradition, and particularly through the writings of Aleister Crowley, who claimed that the creation of the universe began with nothingness, then divided into two, an idea he expressed in the pseudo-mathematical formula "0=2." I would caution, however, against any reading that understands this myth as straightforwardly "monotheist." Miria, the Other, is taken seriously as a separate entity who is yet a part of the original substance (or nothingness—an ambiguity that reflects one of the Western mystery tradition's central paradoxes). For further explication of these terms in a Pagan theological context, see Christine Hoff Kraemer, "Introduction: Pagan Theologies," *Seeking the Mystery: An Introduction to Pagan Theologies* (Englewood, CO: Patheos Press, 2012) Chapter 1 and Aleister Crowley, "The Universe: The 0 = 2 Equation," *Magick without Tears* (Tucson, AZ: New Falcon, 1991) 52–63, http://hermetic.com/crowley/magick-without-tears/mwt_05.html.
11. *The Book of the Law* is constructed as scripture, and Crowley believed that it was dictated to him by a spiritual being while he was in a trance state. Although the text shows the clear influence of Crowley's style, many contemporary Pagans and practitioners of Western esotericism treat it as an inspired work, though not necessarily an infallible one. Aleister Crowley, *The Book of the Law* (1904, 1938; Boston: Weiser Books, 1976) I.29.
12. Starhawk, *Spiral Dance* 48.
13. Starhawk, *Spiral Dance* 48.
14. T. Thorn Coyle, personal communication, drawing on oral tradition from poet and Craft teacher Victor Anderson. Coyle has adopted the term "God Hirself" to emphasize the all-gendered nature of this deity, though I myself prefer the way "God Herself" contains an apparently masculine and apparently feminine term and so emphasizes a state of being multiply-gendered.
15. See Yvonne Aburrow, "Is It Meaningful to Speak of 'Queer Spirituality?' An Examination of Queer and LGBT Imagery and Themes in Contemporary Paganism and Christianity," *Contemporary Christianity and LGBT Sexualities*, Ed. Stephen Hunt (Burlington, VT: Ashgate, 2009) 148.
16. Elaine Pagels, *The Gnostic Gospels* (1979; New York: Vintage Books, 1989) 56.
17. Starhawk, *Spiral Dance* 20–21.
18. Briefly, Western esotericism is a belief system and practice that evolved from a combination of natural philosophy and Jewish and Christian mysticism. Today, it continues to function as

a religiously-inflected system used to refine the self and move into deeper relationship with the spiritual. Western esotericism has been one important influence on the contemporary Pagan movement, which has eclectic roots including pre-Christian indigenous religions, various earth-based spiritualities, and the counterculture of the 1960s. For a history that deals with all of these threads, see Ronald Hutton, *The Triumph of the Moon: A History of Modern Pagan Witchcraft* (Oxford: Oxford UP, 1999).

19. See Jason Lawton Winslade, "When Pan Met Babalon: Challenging Sex Roles at a Thelemic/Pagan Festival," Pagan Studies Group, Annual Meeting of the American Academy of Religion, Chicago, IL, November 16–29, 2012 and Hugh B. Urban, *Magia Sexualis: Sex, Magic, and Liberation in Modern Western Esotericism* (Berkeley, CA: University of California Press, 2006) 257. See also the extensive online discussion of quotes from Crowley's writing entitled, "Aleister Crowley: A Legacy of Sexism/Was Aleister Crowley a Sexist?," Eds. Catherine Yronwode et al., 1999, http://arcane-archive.org/faqs/crowleysexistfaq.php#sexist.
20. See Winslade, "When Pan Met Babalon." Periodic Women's Symposiums hosted by the Ordo Templi Orientis (Crowley's fraternal order) have also addressed a variety of topics related to feminism, Thelema, and the free expression of sexuality. For examples, see "OTO Women's Symposium 2006," NOTOCON.org, 2006, http://notocon.org/past/otows/i/presentations.html, and the work of Thelemic feminist Brandy Williams (http://brandywilliams.org/, author of *The Woman Magician* [2011]).
21. See, for example, Moore's poetic history of same-sex relationships with illustrator José Villarrubia, *The Mirror of Love* (Atlanta, GA: Top Shelf Productions, 2004).
22. For more substantial discussions of the *Promethea* series in relationship to contemporary Paganism and its theology, see Christine Hoff Kraemer and J. Lawton Winslade, "The Magic Circus of the Mind: Alan Moore's *Promethea* and the Transformation of Consciousness through Comics," *Graven Images: Religion in Comic Books and Graphic Novels*, Eds. A. David Lewis and Christine Hoff Kraemer (New York: Continuum, 2010) 274–291 and Kraemer, "The Undying Fire: Divine Love and the Erotic in *Promethea*," *Sexual Ideology in the Works of Alan Moore*, Eds. Todd Comer and Joseph Sommers (Jefferson, NC: McFarland, 2012) 150–162.
23. Here, Moore follows twentieth-century magician and esotericist Aleister Crowley, for whom the Whore of Babylon (or Babalon) was an important religious symbol. For more on the influence of Crowley on Moore, see Kraemer, "The Undying Fire," 150–162.
24. *Promethea* #21. Alan Moore, J.H. Williams (penciller), and Mick Gray (inker), *Promethea* Vol. 4, reprinting issues #19–25 (La Jolla, CA: America's Best Comics, 2003), unpaginated.
25. Loraine Hutchins, "Bisexual Women as Emblematic Sexual Healers and the Problematics of the Embodied Sacred Whore," 2002, *Journal of Bisexuality* 10 (2010): 201–202.
26. Hutchins 202.
27. Alan Moore (writer) and Dave Gibbons (penciller), *Watchmen* (New York: DC Comics, 1987) IX.26–28.
28. Deane Juhan, *Touched by the Goddess: The Physical, Psychological, and Spiritual Powers of Bodywork* (Barrytown, NY: Station Hill of Barrytown, 2002) 30–31.
29. Juhan 31. The notion of "armoring" comes from Reichian psychoanalysis, which is discussed briefly in Chapter 4.
30. Addictive behavior can also focus on numbing or otherwise coping with pain and negative emotions. The topic of addiction is too complex to adequately summarize here, however, so I refer the reader to the enormous body of psychological literature on the topic, particularly in the area of behavioral addiction.
31. Juhan, *Touched* 31–32.

32. Doreen Valiente, "The Charge of the Goddess," The Doreen Valiente Foundation, http://doreenvaliente.org/2009/06/poem-the-charge-of-the-goddess/.
33. Marvin M. Ellison, *Erotic Justice: A Liberating Ethic of Sexuality* (Louisville, KY: Westminster John Knox Press, 1996) 118–119.

2. Politics in the Pit of the Stomach: Primal Pleasures and the Health of the Social Body

1. At the beginning of *Bodymind* (New York: Penguin, 1977, 1986), Ken Dychtwald describes the experience of having his life history read from his body in astonishing detail by body-centered psychotherapist John Pierrakos. *Bodymind* offers an elaborate and highly specific system for correlating personality traits with posture and body shape. As I feel that there are frequently multiple causes for postural deviations and movement habits, however, I find Dychtwald's system to be overly rigid, though its principles are useful as guidelines. Dychtwald was likely influenced by the mid-twentieth century work of psychoanalyst Wilhelm Reich, whom I discuss briefly in Chapter 4. As Robert C. Fuller explains,

   > Believing that the body and mind work together as one whole, [Reich] reasoned that all of our physical expressions reveal something about our personality. Our posture, glances, or even our gait all convey our characteristic personality traits. Thus, for example, persons with chronically stooped shoulders might be keeping their face and eyes hidden as a way of protecting themselves from hurtful contact with others. (137)

   See Robert C. Fuller, *Spirituality in the Flesh: Bodily Sources of Religious Experience* (New York: Oxford University Press, 2008) and Wilhelm Reich, *Character-Analysis* (New York: Noonday, 1949).
2. Deane Juhan, *Touched by the Goddess: The Physical, Psychological, and Spiritual Powers of Bodywork* (Barrytown, NY: Station Hill Press, 2002) 15–16.
3. Juhan 16.
4. Marvin M. Ellison, *Erotic Justice: A Liberating Ethic of Sexuality* (Louisville, KY: Westminster John Knox Press, 1996) 15.
5. Ellison 8.
6. Ellison 9.
7. Ellison 9.
8. Ellison 41. See also Gayle Rubin, "The Traffic in Women: Notes on the 'Political Economy' of Sex," 1975, *Deviations: A Gayle Rubin Reader* (Durham, NC: Duke University Press, 2011) 33–65.
9. Ellison 46.
10. Ellison 48.
11. Ellison 54–55.
12. Ellison 55.
13. Ellison 29.
14. See, for example, Marcella Althaus-Reid, *The Queer God* (New York: Routledge, 2003) 41, 52.
15. By "prophetic," I am referring to the tradition of biblical prophets, who frequently position themselves against those in power in order to speak on behalf of God and the oppressed. Isaiah is one such prophet; Jesus can be seen as another.
16. Ellison 56.

17. Ellison 79, quoting Beverly Wildung Harrison, "Human Sexuality and Mutuality," in *Christian Feminism: Visions of a New Humanity*, Ed. Judith L. Weidman (San Francisco, CA: Harper & Row, 1984) 147–148.
18. Ellison 81.
19. Ellison 81–82.
20. Ellison 86.
21. Ellison 89.
22. See Riane Eisler, *The Chalice and the Blade* (San Francisco, CA: HarperOne, 1988) and *Sacred Pleasure: Sex, Myth, and the Politics of the Body* (San Francisco, CA: HarperSanFrancisco, 1996).
23. See Cynthia Eller, *The Myth of Matriarchal Prehistory: Why an Invented Past Won't Give Women a Future* (Boston, MA: Beacon Press, 2000).
24. Dale Peterson and Richard Wrangham, *Demonic Males: Apes and the Origins of Human Violence* (New York: Houghton Mifflin, 1997); "Chimps and Bonobos," *Evolution*, PBS.org, 2001, www.pbs.org/wgbh/evolution/library/07/3/l_073_03.html.
25. Riane Eisler, *Sacred Pleasure: Sex, Myth, and the Politics of the Body* (San Francisco, CA: HarperSanFrancisco, 1996) 217.
26. Dossie Easton and Janet W. Hardy, *The New Topping Book* (Emeryville, CA: Greenery Press, 2003) 177–192 and *Radical Ecstasy* (Oakland, CA: Greenery Press, 2004) 181–183.
27. Rubin, "Thinking Sex" 176.
28. "Head Injuries in Football," *New York Times*, October 21, 2010, www.nytimes.com/info/concussions-in-football/.
29. Chas S. Clifton, *Her Hidden Children* (Lanham, MD: AltaMira Press, 2006) 58–59.
30. Clifton 65.
31. Peterson and Wrangham 69, 132.
32. Nancy Eisenberg and Janet Strayer (Eds.), *Empathy and its Development* (New York: Cambridge University Press, 1990) 5–7.
33. Stacey Kennelly, "Educating for Empathy," *Greater Good: The Science of a Meaningful Life*, July 18, 2012, http://greatergood.berkeley.edu/article/item/educating_for_empathy.
34. Jill Suttie, "The Politics of Empathy," *Greater Good: The Science of a Meaningful Life*, June 16, 2010, http://greatergood.berkeley.edu/article/item/the_politics_of_empathy. Suttie profiles Jeremy Rifkin's *The Empathic Civilization: The Race to Global Consciousness in a World in Crisis* (Penguin, 2009) and Frans De Waal's *The Age of Empathy: Nature's Lessons for a Kinder Society* (Harmony Books, 2009).
35. Despite his Christian affiliations, Fox is sympathetic to and actively in conversation with contemporary Paganism and was interviewed for the book *Modern Pagans*, Eds. V. Vale and John Sulak (San Francisco, CA: Re/Search, 2001) 116–118.
36. See Starhawk, *The Fifth Sacred Thing* (New York: Bantam Books, 1993) and Riane Eisler, *The Chalice and the Blade* (San Francisco, CA: HarperOne, 1988).
37. Some players do have "24/7" or full-time dominance/submission relationships, a marriage-like commitment that may be renegotiated periodically or annually. At least on the surface, these BDSM relationships threaten to entrench internal power hierarchies to a much greater degree than those where power exchange is more dynamic. 24/7 relationships are relatively rare, and I do not feel I understand their dynamics well enough to recommend or condemn them. However, those in 24/7 BDSM relationships are usually much more deliberate about their power exchange than the average married couple. A couple who have consciously taken on permanent roles of top and bottom may be better equipped to treat each other with respect than a married couple with a similarly large power differential (for example, where one partner is older and has significant financial resources and social status, and the other is

younger, uneducated, and in debt). In my experience, 24/7 D/s relationships are often invisible to observers, and advocates emphasize the importance of both partners treating the other with respect and supporting the other in his personal development. Pagan practitioners Raven Kaldera and his partner Joshua Tenpenny, for example, have written about 24/7 relationships as a spiritual discipline in several books, including *Sacred Power, Holy Surrender: Living a Spiritual Power Dynamic (*Hubbardston, MA: Alfred Press, 2010) and *Power Circuits: Polyamory in a Power Dynamic* (Hubbardston, MA: Alfred Press, 2010).

38. Ellison 87.
39. Carter Heyward, *Touching Our Strength: The Erotic as Power and the Love of God* (San Francisco, CA: Harper & Row, 1989) 107–108.
40. Heyward 108–109.
41. Dossie Easton and Janet W. Hardy, *The New Bottoming Book* (Emeryville, CA: Greenery Press, 2001) 167.
42. Easton and Hardy, *Bottoming* 24–25.
43. Starhawk, *Truth or Dare: Encounters with Power, Authority, and Mystery* (San Francisco, CA: HarperSanFrancisco, 1987) 9.
44. M. Macha NightMare, "Reclaiming-Tradition Witchcraft," *Reclaiming Quarterly* #88 (Autumn 2002), www.reclaimingquarterly.org/88/rq-88-tradition.html.
45. M. Macha NightMare, "Back Story, One of Several," *Broomstick Chronicles*, August 8, 2012, www.besom.blogspot.com/2012/08/back-story-one-of-several.html.
46. M. Macha NightMare, "A Co-Founder Withdraws from Reclaiming Tradition," *Broomstick Chronicles*, August 6, 2012, www.besom.blogspot.com/2012/08/a-co-founder-withdraws-from-reclaiming.html.
47. Anne Hill, *The Baby and the Bathwater: What I Learned About Spirituality, Magic, Community, Ecstasy and Power from 25 Years in Reclaiming* (Bodega Bay, CA: Serpentine Media & Music, 2012)
48. Personal communication, with acknowledgment of the work of Hannah Arendt.

3. *Erotic Eclecticism and Divine Deviance*

1. Gayle Rubin, "Thinking Sex: Notes for a Radical Theory of the Politics of Sexuality," 1984, *Deviations: A Gayle Rubin Reader* (Durham, NC: Duke University Press, 2011) 155.
2. Rubin, "Thinking Sex" 181.
3. For more on contemporary Pagan virtue ethics, see Christine Hoff Kraemer, *Seeking the Mystery: An Introduction to Pagan Theologies* (Englewood, CO: Patheos Press, 2012) Chapter 5; Brendan Myers, *The Other Side of Virtue* (Hants, UK: O Books, 2008); and Ellen Friedman, *As Above, So Below: A System of Value-Based Ethics for Wiccan Clergy*, 2001 (Seattle, WA: City University, 2011).
4. Rubin, "Thinking Sex" 154.
5. Foucault argues further that the pleasure gained from monitoring, controlling, and eliciting confessions from patients, as well as the pleasure of evading, transgressing, submitting to, and then escaping such control, were also underlying motivations for both professionals and patients participating in this medical system. Although Foucault does not use the vocabulary of BDSM explicitly, his observations of this system of semi- or non-consensual erotic power exchange provide one historical model of how erotic energy can become twisted under oppressive social conditions. See *History of Sexuality Vol. 1*, trans. Robert Hurley (1978; New York: Vintage Books, 1990) 36–74.

6. See Foucault 43, 101. David Halperin also develops and defends this point at greater length in David Halperin, *One Hundred Years of Homosexuality: And Other Essays on Greek Love* (New York: Routledge, 1990).
7. Rubin, "Thinking Sex" 155.
8. Melissa M. Wilcox, "Innovation in Exile: Religion and Spirituality in Lesbian, Gay, Bisexual, and Transgender Communities," *Sexuality and the World's Religions*, Eds. David W. Machacek and Melissa M. Wilcox (Santa Barbara, CA: ABC-CLIO, 2003) 323–358.
9. Eve Kosofsky Sedgwick, *Tendencies* (Durham: Duke UP, 1993) xii.
10. Althaus-Reid was unable to attend due to illness; her paper "The Church as Dis/Grace" was read by Mark Jordan. *Queering the Church: Changing Ecclesial Structures*, Center for Practical Theology, Boston University School of Theology (Boston, MA, April 19, 2007).
11. Pat Califia, "A Personal View of the History of the Lesbian S/M Community and Movement in San Francisco," *Coming to Power*, Ed. SAMOIS, Second Edition (Boston, MA: Alyson Books: 1982) 245–283.
12. Sadomasochism, or S/M, is an alternative term for BDSM, an acronym for "bondage and discipline/dominance and submission/sadomasochism." It is used here because it is the preferred term of the writers being discussed.
13. Califia, "Personal View" 255–256.
14. Califia, "Personal View" 256–257.
15. Recent evidence even suggests that consuming pornography may reduce the incidence of rape. Todd Kendall argued in a talk given at Stanford University that the greater availability of pornography due to the Internet may actually be reducing the incidence of rape, especially among young men aged 15–19: "Pornography, Rape, and The Internet," September 28, 2006, www.law.stanford.edu/display/images/dynamic/events_media/Kendall%20cover%20+%20pap er.pdf. See also Pat Califia, *Public Sex: The Culture of Radical Sex* (San Francisco, CA: Cleis Press, 1994) 100, referencing the *Attorney General's Commission on Pornography, Final Report*, July 1986, Vol. I, 198–199, 206.
16. For example, the 2004 documentary *Thinking XXX* accompanies photographer Timothy Greenfield-Sanders' book of clothed and nude portraits of porn stars. Through informal interviews of the actors, the documentary is both critical of the porn industry and demonstrates ways that some porn stars have found fulfillment and meaning in their work.
17. Rubin, "Thinking Sex" 168.
18. See Rubin's note to the 1992 revision of "Thinking Sex" 183.
19. Gayle Rubin, "Elegy for the Valley of Kings," In *Changing Times: Gay Men and Lesbians Encounter HIV/AIDS*, Eds. Martin P. Levine et al. (Chicago, IL: University of Chicago Press, 1997) 101–144.
20. Rubin, "Thinking Sex" 169. Although I attempted to find sex crime statistics that compared BDSM practitioners to non-BDSM practitioners, I was unable to find either studies or anecdotal evidence addressing this topic. It is, of course, always impossible to *prove* the absence of a connection. Considering the low regard that BDSM practice is assigned in our culture and the well-publicized hostility of second-wave feminism to it, however, I must conclude that, as Rubin claims, no statistical evidence exists to link BDSM practice with violent sex crime.
21. Rubin, "Thinking Sex" 169.
22. "Facts about Violence," *Feminist.com*, 2008, www.feminist.com/antiviolence/facts.html.
23. Rubin, "Thinking Sex" 151.
24. See, for example, Marty Klein and Charles Moser, "SM (Sadomasochistic) Interests as an Issue in a Child Custody Proceeding," *Journal of Homosexuality* 50: 2 (2006): 233–242, http://sexresearch.com/SSSS/PDFs/20060502_Klein_Moser.pdf.

25. Juliet Richters et al. state, "Our findings support the idea that BDSM is simply a sexual interest or subculture attractive to a minority, and for most participants not a pathological symptom of past abuse or difficulty with 'normal' sex." See "Demographic and Psychosocial Features of Participants in Bondage and Discipline, 'Sadomasochism' or Dominance and Submission (BDSM): Data from a National Survey," *Journal of Sexual Medicine* 5: 7 (March 2008): 1660–1668, www3.interscience.wiley.com/journal/120126317/abstract. See also Gabriele Hoff & Richard A. Sprott, "Therapy Experiences of Clients with BDSM Sexualities: Listening to a Stigmatized Sexuality," *Electronic Journal of Human Sexuality* 12 (September 2009), www.ejhs.org/Volume12/bdsm.htm and Margaret Nichols, "Psychology and BDSM: Pathology or Individual Difference?" *New Jersey Psychologist* (2002), www.ipgcounseling.com/psychology_and_bdsm.html.
26. See, for example, Dan Savage, "Youth Pastor Watch," *The Stranger*, June 28, 2010, http://slog.thestranger.com/slog/archives/2010/06/28/youth-pastor-watch. A site search for "Youth Pastor Watch" returns a series of such articles.
27. Dan Gilgoff, "Catholic Church's sex abuse scandal goes global," *CNN World*, March 19, 2010, www.cnn.com/2010/WORLD/europe/03/19/catholic.church.abuse/index.html.
28. Dossie Easton and Janet W. Hardy, *Radical Ecstasy* (Oakland, CA: Greenery Press, 2004) 56.
29. Jennifer Hunter, *Rites of Pleasure: Sexuality in Wicca and NeoPaganism* (New York: Citadel Press, 2004) 152.
30. Rubin, "Thinking Sex" 154.
31. Jason Pitzl-Waters, "Interview with Starhawk," *The Wild Hunt*, September 28, 2009, http://wildhunt.org/2009/09/interview-with-starhawk.html.
32. Riane Eisler, *The Chalice and the Blade* (San Francisco, CA: HarperOne, 1988).
33. John Michael Greer, "Myth, History, and Pagan Origins," *The Pomegranate: The International Journal of Pagan Studies* 9 (1999): 49.
34. Greer 50.
35. Easton and Hardy, *Radical Ecstasy* 24–25.
36. Robert C. Fuller, *Spirituality in the Flesh: Bodily Sources of Religious Experience* (New York: Oxford University Press, 2008) 134, summarizing Elaine Scarry, *The Body in Pain: The Making and Unmaking of the World* (New York: Oxford University Press, 1985).
37. Fuller 135–136, summarizing Ariel Glucklich, *Sacred Pain: Hurting the Body for the Sake of the Soul* (New York: Oxford University Press, 2001).
38. Raven Kaldera, *Hermaphrodeities*, Second Edition (Hubbardston, MA: Asphodel Press, 2009) 11–15.
39. Cynthia Eller, *Am I A Woman? A Skeptic's Guide to Gender* (Boston, MA: Beacon Press, 2003) 20–21.
40. Eller 24.
41. Kaldera, *Hermaphrodeities* 21.
42. Kate Bornstein, *Gender Outlaw* (New York: Vintage Books, 1995) 65–66.
43. Statistics are available from Marti Abernathey, "International Transgender Day of Remembrance: Statistics and Other Info," last update November 17, 2012, www.transgenderdor.org/statistics.
44. See Simon LeVay, *Gay, Straight, and the Reason Why: The Science of Sexual Orientation* (Oxford: Oxford University Press, 2012).
45. Constance Wise, *Hidden Circles in the Web: Feminist Wicca, Occult Knowledge, and Process Thought* (Lanham, MD: Altamira Press, 2008) 67.
46. Wise 68.

47. For more on gender debates in contemporary Paganism, see Christine Hoff Kraemer, "Gender and Sexuality in Contemporary Paganism," *Religion Compass* 6:8 (2012): 390–401.
48. The case of the Divilbiss family is perhaps best-known in the ethical non-monogamy community. In 1998, the grandparents of a child successfully challenged a polyamorous mother's custody. After fighting the ruling for several years, April Divilbiss eventually relented, acknowledging that she had been financially unable to provide a safe and stable environment for her daughter, and that this was her mother-in-law's true reason for attempting to gain custody. Nevertheless, Divilbiss' appearance on television as a member of a polyamorous family was the primary evidence that caused the judge to rule that she was an unfit parent. See Woody Baird, "Pagan Mother Battles for Custody," *PolyamorySociety.org*, January 11, 1999, www.polyamorysociety.org/Yahoo-Divilbiss_Article.html and April Divilbiss, "PolyFamily Child Custody Case Ends After 2 Year Battle," *PolyamorySociety.org* 1998, www.polyamorysociety.org/Divilbiss_Families_Case_Ends.html.
49. Arno Karlen, *Threesomes: Studies in Sex, Power, and Intimacy* (New York: Beech Tree Books, 1998) 304–312.
50. Raven Kaldera, *Pagan Polyamory: Becoming a Tribe of Hearts* (Woodbury, MN: Llewellyn Publications, 2005) xxii.
51. Kaldera 177.
52. John Humphrey Noyes, *First Annual Report of the Oneida Association* (Oneida Reserve, NY: Leonard, 1849) 25, 27 [original numbering removed]; quoted in Foster, *Religion 3*. Notably, however, the Oneida community's idea of "free love" was actually highly structured, and Noyes's interest in eugenics has sometimes been seen as "prefiguring some of the most repressive and threatening human engineering experiments of the twentieth century" (Foster, "Free Love" 253). See Lawrence Foster, "Free Love and Community: John Humphrey Noyes and the Oneida Perfectionists," *America's Communal Utopias*, Ed. Donald E. Pitzer (Chapel Hill: University of North Carolina Press, 1997) 253–278. A potentially revealing comparison might be made between the rule-boundedness of some contemporary polyamorists (a minority of whom organize their sexual relationships through lengthy written contracts) and the complex marriage of the Oneida community.
53. Dossie Easton and Janet Hardy (writing as Catherine. A Liszt), *The Ethical Slut: A Guide to Infinite Sexual Possibilities* (San Francisco, CA: Greenery Press, 1997); Deborah M. Anapol, *Polyamory: The New Love without Limits* (San Rafael, CA: IntiNet Resource Center, 1997).
54. Tristan Taormino, *Opening Up: A Guide to Creating and Sustaining Open Relationships* (San Francisco, CA: Cleis Press, 2008).
55. "Good Morning America: Oral Sex is the New Goodnight Kiss," YouTube, May 28, 2009, www.youtube.com/watch?v=Yj8p94a81ZE.
56. Sarah Knoploh, "ABC Exposes Prevalence of Oral Sex among Teen Girls," *Culture and Media Institute*, May 28, 2009, www.cultureandmediainstitute.org/about/about.aspx.
57. See Christopher Trenholm, Barbara Devaney, Ken Fortson, Lisa Quay, Justin Wheeler, and Melissa Clark, "Impacts of Four Title V, Section 510 Abstinence Education Programs, Final Report," Mathematica Policy Research (April 2007), http://aspe.hhs.gov/hsp/abstinence07/index.htm, and The Foundation for AIDS Research, "Assessing the Efficacy of Abstinence-Only Programs for HIV Prevention among Young People," Issue Brief No. 2 (Revised October 2007), http://thehill.com/images/stories/whitepapers/pdf/AbstinenceIB.pdf.
58. Deborah L. Tolman, *Dilemmas of Desire: Teenage Girls Talk about Sexuality* (Cambridge, MA: Harvard University Press, 2002) 15.
59. Tolman 3.

60. James R. Kincaid, *Erotic Innocence: The Culture of Child Molesting* (Durham, NC: Duke University Press, 1998) 31. See also 13, 18–19, 30–50.
61. Califia, *Public Sex* 67.
62. Phoebe Gloeckner, *The Diary of a Teenage Girl: An Account in Words and Pictures* (Berkeley, CA: Frog, 2002) and Craig Thompson, *Blankets* (Marietta, GA: Top Shelf, 2003).
63. Christine Hoff Kraemer, T*he Erotic Fringe: Sexual Minorities and Religion in Contemporary American Literature and Film* (Ph.D. Boston University, 2008) 247, 285.
64. Heather Corinna, "An Immodest Proposal," In *Yes Means Yes! Visions of Female Sexual Power and a World without Rape*, Eds. Jaclyn Friedman and Jessica Valenti (Berkeley, CA: Seal Press, 2008) 187.
65. Bornstein 97–98.

4. *The Sacrament of Touch*

1. In a Pagan Studies panel on the meanings of "idolatry" at the 2009 American Academy of Religion conference in Montréal, Quebec, Canada, Michael York expressed a preference for the term "holy" above the term "sacred." The root meaning of "holy" refers to wholeness and health, while that of "sacred" refers to a special separation, a setting-apart.

   Etymologically, "holy" reflects contemporary Paganism's emphasis on imminent divinity much more accurately than "sacred," which implies that the divine must be separated from the profane, mundane world. Contemporary Pagans writers, however, have largely preferred the term "sacred" to "holy," possibly because "holy" is so frequently used in Christian liturgy and hymns ("Holy Spirit," etc.). Although I am sympathetic to York's argument and so have chosen to use "holy" to describe those things that awaken the divine in human beings, the historical connection of "sacrament" with unearned divine love makes the term appropriate here, despite its etymological roots.
2. *The 1928 Book of Common Prayer* (New York: Oxford University Press, 2007) 292.
3. Those unfamiliar with the full diversity of Christian traditions may assume that the concept of "grace" requires a belief in original sin. Not all Christian theologians have adhered to Augustine's doctrine, however—and contemporary Pagans may find much to admire in the writings of Augustine's fourth-century opponent Pelagius, whose earth-centered theology inspired Matthew Fox's contemporary theology of original blessing. (Perhaps significantly, before being excommunicated from the Catholic Church, Fox's theology was condemned by church authorities as "pagan heresy.") See Matthew Fox, *Original Blessing* (Santa Fe, NM: Bear & Co., 1983) and Martin Wroe, "Turbulent priest ministers to the New Age soul: Martin Wroe meets the friar delivering 'pagan heresy' to packed churches," *The Independent*, July 14, 1992, www.independent.co.uk/news/uk/turbulent-priest-ministers-to-the-new-age-soul-martin-wroe-meets-the-friar-delivering-pagan-heresy-to-packed-churches-1533115.html.
4. In considering how to speak of touch as a religious practice in this chapter, I was also inspired by the Jewish concept of *mitzvah*, which has connotations of both good deed and holy obligation. Those who are conversant with the Jewish tradition may find this connection adds an additional layer of meaning to my explanation of touch as a sacrament.
5. Tiffany Field, *Touch* (Cambridge, MA: MIT Press) 13–14; Robert Noah Calvert, *The History of Massage: An Illustrated Survey* (Rochester, VT: Healing Arts/Inner Traditions, 2002) 1.
6. Calvert 14, 45.
7. Calvert 12–61.
8. Catherine L. Albanese, *Nature Religion in America* (Chicago, IL: University of Chicago Press, 1991) 123.

9. See Robert C. Fuller, *Mesmerism and the American Care of Souls* (Philadelphia, PA: University of Pennsylvania Press, 1982).
10. Similar notions, in fact, can be found at least as far back as the writings of sixteenth-century natural philosopher Paracelsus. See Chas Clifton, *Her Hidden Children* (Lanham, MD: AltaMira Press, 2006) 63.
11. David Dunér, *The Natural Philosophy of Emanuel Swedenborg* (New York: Springer, 2013) 364; Albanese 122.
12. Albanese 82. Albanese also points out, however, that this view in American Transcendentalism is in tension with a philosophical idealism that sees matter as ultimately illusory, especially in the writings of Ralph Waldo Emerson.
13. Albanese 132–146.
14. Clifton 62. Reich's work has been carried on in newly developed body-oriented therapies such as Alexander Lowen's bioenergetics, which views the body's ability to make pleasurable responses to the environment as the primary measure of health. See Robert C. Fuller, *Spirituality in the Flesh: Bodily Sources of Religious Experience* (New York: Oxford University Press, 2008) 138.
15. Clifton 62.
16. Urban 112–113.
17. Urban 11.
18. Ronald Hutton, *Witches, Druids, and King Arthur* (London: Hambledon & London, 2003) 193–214. In Wicca, ritual nudity is known as "going skyclad."
19. Clifton 64. Clifton mentions the 1930s Church of Aphrodite, the Church of All Worlds, Feraferia, and the Anderson Faery (or Feri) tradition as groups that approach sex sacramentally.
20. See Stephanie Lynn Budin, *The Myth of Sacred Prostitution in Antiquity* (Cambridge: Cambridge University Press, 2008).
21. See Ellen Evert Hopman and Lawrence Bond (Eds.), "Sacred Prostitutes," *People of the Earth: The New Pagans Speak Out* (Rochester, VT: Destiny Books, 1996) 139–151. This chapter features interviews with D'vora and Annie Sprinkle. For the problematics of "sacred prostitute" imagery in the context of sex work, see Loraine Hutchins, "Bisexual Women as Emblematic Sexual Healers and the Problematics of the Embodied Sacred Whore," 2002, *Journal of Bisexuality* 10 (2010): 336–357.
22. Sarah Pike, "Gleanings from the Field: Leftover Tales of Grief and Desire," *Researching Paganisms*, Eds. Jenny Blain, Douglas Ezzy, and Graham Harvey (Walnut Creek, CA: AltaMira Press, 2004) 105.
23. See J. Gordon Melton, "Reiki: The International Spread of a New Age Healing Movement," *New Age Religion and Globalization*, Ed. Mikael Rothstein (Aarhus: Aarhus University Press, 2001) 73–93.
24. Similar, but somewhat less common forms of contemporary energy healing or healing touch include therapeutic touch, polarity therapy, and energy-field healing. For a summary of healing touch practices in contemporary American religion, see Candy Gunther Brown, "Touch and American Religions," *Religion Compass* 3/4 (2009): 770–783.
25. Deane Juhan, *Touched by the Goddess: The Physical, Psychological, and Spiritual Powers of Bodywork* (Barrytown, NY: Station Hill Press, 2002) 65.
26. Deane Juhan, *Job's Body: A Handbook for Bodywork*, Third Edition (Barrytown, NY: Station Hill Press, 2003 [1987]) 3.
27. David Lauterstein, *The Deep Massage Book* (Taos, NM: Complimentary Medicine Press, 2011) 44–45.

28. Lauterstein 45.
29. Martin Buber, *I and Thou* (New York: Scribner, 1958).
30. Phyllis K. Davis, *The Power of Touch*, Revised Edition (Carlsbad, CA: Hay House, 1999 [1991]) 80.
31. Juhan, *Job's Body* 44, referring to Gardner, I.L., referring to Spitz, R.A. "Hospitalism: A follow-up report." In D. Fenichel, P. Greenacre & A. Freud (Eds.), *The Psychoanalytic Study of the Child, Vol. 2*, New York: International Universities Press, 1947, 113–117. See also Montagu, *Touching* Chapter. 4.
32. Ashley Montagu, *Touching: The Human Significance of the Skin*, Third Edition (New York: Harper & Row, 1986 [1971]) 99.
33. Juhan, *Job's Body* 47, referring to Montagu.
34. Montagu 38–41. See also H.F. Harlow and R. Zimmerman, "Affectional Responses in the Infant Monkey," *Science* 130 (1959): 421 and H.F. Harlow, "The Nature of Love," *The American Psychologist* 13 (1958): 673–685.
35. Davis 108, citing Harlow and Zimmerman.
36. Montagu 28.
37. Montagu 20–21.
38. Montagu 238–239.
39. For additional summary of animal studies that point to the importance of tactile stimulation for physical and behavioral development, see Montagu, *Touching* Chapter 1, "The Mind of the Skin."
40. Montagu 113.
41. Montagu 164–165.
42. In Chapter 8, "Culture and Contact," Montagu cites a variety of anthropological studies of cultures around the world to suggest that such parenting techniques produce more peaceful societies. These summaries are many decades old and lack quantitative data, however. *Touching* is a product of the late 1960s and shows a strong Freudian bias in its anthropology, so I have chosen not to cite these parts of the book as authoritative. The animal and infant studies that Montagu describes, based (as many of them are) on survival rates and other quantitative measurements, retain their relevance despite shifting scholarly attitudes towards race, class, and gender.
43. Davis 109, referring to James W. Prescott, "Body Pleasure and the Origins of Violence," *The Futurist* April 1975 and Prescott, "Alienation of Affection," *Psychology Today* December 1979. See also J.W. Prescott, "Affectional bonding for the prevention of violent behaviors: Neurobiological, psychological and religious/spiritual determinants," *Violent Behavior: Assessment and Intervention*, Vol. 1, Eds. L.J. Herzberg, G.F. Ostrum, & J. Roberts Field (Great Neck, NY: PMA Publishing, 1990).
44. Davis 111.
45. Tiffany Field, *Touch* (Cambridge, MA: MIT Press) 61.
46. Shirley Vanderbilt, "Eating Disorders: Touch Can Ease the Symptoms," *Massage and Bodywork Magazine*, August/September 2001, www.massagetherapy.com/articles/index.php/article_id/552/Eating-Disorders. See also Field, *Touch* 137.
47. See, for example, Davis 72, referencing Mimi Cruz, "No Hugs? Get a Grip, Kids Say," *Los Angeles Times*, February 12, 1998, B1; Sarah Harris, "Teachers ordered not to risk touching children for fear of legal action," *Daily Mail*, February 25, 2008, www.dailymail.co.uk/news/article-518289/Teachers-ordered-risk-touching-children-fear-legal-action.html; Maria Newman, "Cautious Teachers Reluctantly Touch Less; A Fear of Abuse Charges Leads to Greater Restraint With Students," *New York Times*, June 24, 2008, www.nytimes.com/1998/

06/24/nyregion/cautious-teachers-reluctantly-touch-less-fear-abuse-charges-leads-greater.html; and Alyssa Newcomb, "New Jersey School Bans Hugging," *ABC News*, March 23, 2012, http://abcnews.go.com/blogs/headlines/2012/03/new-jersey-school-bans-hugging/.

48. Tiffany Field, "American Adolescents Touch Each Other Less and Are More Aggressive toward Their Peers as Compared with French Adolescents," *Adolescence* 34: 136 (Winter 1999): 753–758, http://findarticles.com/p/articles/mi_m2248/is_136_34/ai_59810232/.
49. Ironically, such strictures around touch arose from Victorians' belief in the "omnipresence and power of the sex instinct," as Steven Seidman puts it: "The Victorians believed that arousing sexual feelings easily elicits sensuality; sex threatened to engulf marriage in a sea of lust. To preserve the spiritual essence of marriage, the Victorians felt compelled to desensualize sex or to erect elaborate barriers to contain erotic desires" (49). See Seidman, "The Power of Desire and the Danger of Pleasure: Victorian Sexuality Re-Considered," *Journal of Social History* 24: 1 (Autumn, 1990): 47–67.

    Perhaps partially in reaction to such strictures, religious subcultures focused on healing through physical touch also thrived during the Victorian era, as did sexual mysticism, prostitution, and an underground trade in pornographic and erotic art. See Catherine L. Albanese, *Nature Religion in America* (Chicago, IL: University of Chicago Press, 1991) 117–152; Hugh B. Urban, *Magia Sexualis: Sex, Magic, and Liberation in Modern Western Esotericism* (Berkeley, CA: University of California Press, 2006) 55–139; and Steven Marcus, *The Other Victorians: A Study of Sexuality and Pornography in Mid-Nineteenth-Century England*, 1966 (Piscataway, NJ: Transaction Publishers, 2008).
50. For example, as Michel Foucault describes in *The History of Sexuality*, the medical culture of the Victorian era systematically classified and pathologized individuals based on their sexual "deviance," leading to the creation of such categories as "homosexual," "masturbator," and "nymphomaniac." See *History of Sexuality Vol. 1*, trans. Robert Hurley (New York: Vintage Books, 1978 [1976]) 42–43.
51. Davis 85–88.
52. Davis 86.
53. Davis 22–23.
54. Juhan, *Touched* 31–32.
55. Field, *Touch* 75–90.
56. Davis 201.
57. Davis 195–209.
58. Juhan, *Job's Body* 18.

5. *A Love Stronger than Fear*

1. Therapeutic and/or spiritually-based erotic healing work is legal in some states, such as California, where practitioners can be trained in a state-certified program in sexological bodywork. (See Association of Certified Sexological Bodyworkers, http://sexologicalbodyworkers.org/.) One recent attempt to practice sacred sex work while using religious nonprofit status as legal protection has failed, however; in late 2011, the Phoenix Goddess Temple of Arizona (www.phoenixgoddesstemple.org/) was raided as an alleged brothel. Practitioners at the temple were providing classes and sexual healing sessions to donors, some of whom donated sums as large as $600 for a single session. Prosecutors condemned the temple as a straightforward prostitution business with a thin veneer of evasive spiritual language. After examining the temple website, my perception is that the temple's intent to offer spiritually based sexual healing is sincere. However, the website is entirely lacking in the descriptions of training in

ethics, communication, trauma counseling, anatomy and physiology, and other concrete tools for therapeutic work that characterize the California's sexological bodywork certification. As a meaningful model for sacred sexual healing work, the Phoenix Goddess Temple leaves much to be desired. See also Christina Caron, "Phoenix Goddess Temple Raided as Alleged Brothel," *ABC News*, September 9, 2011, http://abcnews.go.com/US/phoenix-goddess-temple-raided-alleged-brothel/story?id=14481945#.UVnTEhzqnGE.

2. UniteWomen.org, Facebook.com profile, Timeline Photos, November 29, 2012, www.facebook.com/photo.php?fbid=439038836157661&set=a.302468136481399.68437.302448176483395.
3. Although I am speaking here primarily of communities bound together by social ties and shared erotic practice, sex-positive activism can also take the form of supporting advocacy organizations such as the nonprofit National Coalition for Sexual Freedom (www.ncsfreedom.org/), which provides support for the legal decriminalization of BDSM practices.
4. Brian Alexander, "Incurable Gonorrhea may be next superbug," *Sexploration* on MSNBC.com, April 8, 2010, www.msnbc.msn.com/id/36229547/ns/health-sexual_health/.
5. Don S. Dizon and Michael L. Krychman, *Questions & Answers about Human Papilloma Virus (HPV)* (Sudbury, MA: Jones & Bartlett Publishers, 2011) 10.
6. "Cervical Cancer Statistics," *Centers for Disease Control and Prevention*, November 23, 2010, www.cdc.gov/cancer/cervical/statistics/; National Highway Traffic Safety Administration, "An Analysis of the Significant Decline in Motor Vehicle Traffic Fatalities in 2008," June 2010, p. 1, www-nrd.nhtsa.dot.gov/Pubs/811346.pdf.
7. Deane Juhan, *Touched by the Goddess: The Physical, Psychological, and Spiritual Powers of Bodywork* (Barrytown, NY: Station Hill Press, 2002) 4–5.
8. See, for example, evangelical preacher Jerry Falwell's infamous 1993 statement that "AIDS is not just God's punishment for homosexuals; it is God's punishment for the society that tolerates homosexuals." Bill Press, "The Sad Legacy of Jerry Falwell," *The Milford Daily News*, May 18, 2007, www.milforddailynews.com/opinion/x1987843539.
9. David Schnarch, *Passionate Marriage: Keeping Love & Intimacy Alive in Committed Relationships* (New York: Henry Holt & Company, 1997).
10. Schnarch 91–94.
11. Phyllis K. Davis, *The Power of Touch*, Revised Edition (1991; Carlsbad, CA: Hay House, 1999) 99.
12. San Francisco-based sex writer Carol Queen's Paganism-influenced sex parties, called "Queen of Heaven" parties, are profiled in an issue of the Reclaiming tradition's newsletter. Jennifer Privateer, "Breaking the Boundaries of Sex and Love: An Interview with Madrone," *Reclaiming Quarterly* #82 (Spring 2001), www.reclaimingquarterly.org/82/rq-82-breakingboundaries.html. Loraine Hutchins lists a number of such gatherings, including the Queen of Heaven parties, the annual Body Sacred gathering in New York, and the Body Electric School in San Francisco as examples of functional erotic communities. See Hutchins, "Playing with Sacred Fire: Building Erotic Communities," *Becoming Visible: Counseling Bisexuals across the Lifespan*, Ed. Beth Firestein (New York: Columbia University Press, 2006) 336–357. Similar organizations include the nonprofit Center for Sex Positive Culture in Seattle, Washington, which regularly hosts both ordinary socials and sex and play parties for members.
13. Patrick Califia, *Speaking Sex to Power: The Politics of Queer Sex* (San Francisco, CA: Cleis Press, 2002) 215.
14. Califia 215, referring to Amy Pagnozzi, "Sexual Intolerance Is the Shame in Suicide," *Hartford (Conn.) Courant*, January 29, 2012, http://articles.courant.com/2002–01–29/news/0201291092_1_adult-entertainment-theater-shame.

15. Hutchins 351–352.
16. Brian Alexander, "One preacher's message: have hotter sex," MSNBC.com December 4, 2006 www.msnbc.msn.com/id/13834042/.
17. *Christian Nymphos*, http://christiannymphos.org/.
18. Bob Gifford, "The Virtues of Evangelical Sex," *Cogito, Ergo Dubito* January 25, 2007, http://ergodubito.org/?p=568. In 2006, Haggard lost his job and much of his credibility after prostitute and masseur Mike Jones reported that Haggard had been paying him for sex for three years. Jones explained his reasons for exposing Haggard by saying, "I had to expose the hypocrisy. He is in the position of influence of millions of followers, and he's preaching against gay marriage . . . But behind everybody's back doing what he's preached against." Dan Harris, "Haggard admits buying meth," ABCNews.com, November 3, 2006 http://abcnews.go.com/GMA/story?id=2626067.
19. Hugh B. Urban, *Magia Sexualis: Sex, Magic, and Liberation in Modern Western Esotericism* (Berkeley, CA: University of California Press, 2006) 251, quoting Herbert Marcuse, *Eros and Civilization: A Philosophical Inquiry into Freud* (1955; reprint, Boston, MA: Beacon Press, 1966) xii–xiii.
20. Aleister Crowley, *The Book of the Law* (Boston: Weiser Books, 1976) I.41.
21. Urban 256–257.
22. See Urban, Chapter 6.
23. Hutchins 353.
24. Hutchins 354.
25. Hutchins 353.
26. Fowles 124.
27. Anahita-Gula, quoted in Jennifer Hunter, *Rites of Pleasure: Sexuality in Wicca and NeoPaganism* (New York: Citadel Press, 2004) 108.
28. For example, Patrick Califia notes the persistence of gender-based bigotry even within queer and kink communities in his essay "A Place at the Table," which tells part of the story of his transgender (FTM) partner's decision to become pregnant. Although Califia notes that most of the feedback from their community—people of all genders and sexual orientation—was supportive, some transpeople were openly hostile, and their family configuration was controversial among those who support pursuing civil rights through mainstreaming (45). The impulse for gay activists to disavow those LGBT persons whose practices are perceived as "too radical" is a divisive issue in the gay marriage debate. See Califia, *Speaking Sex to Power* 41–45.
29. Phil Brucato, quoted in Hunter 107.
30. Gayle Rubin, "Thinking Sex: Notes for a Radical Theory of the Politics of Sexuality," 1984, *Deviations: A Gayle Rubin Reader* (Durham, NC: Duke University Press, 2011) 149.
31. Dossie Easton and Janet W. Hardy, *Radical Ecstasy* (Oakland, CA: Greenery Press, 2004) 202–203.
32. Crowley, *The Book of the Law* I.29.
33. Pat Califia, "The Way We Live Now," *Poz Magazine* (May 1999), www.poz.com/articles/214_1517.shtml.

Conclusion: Uncovering the Flesh of the World

1. Associated Press, "Austin City Council Endorses Same-Sex Marriage," *Amarillo Globe News*, September 28, 2012, http://amarillo.com/news/latest-news/2012–09–28/austin-city-council-endorses-same-sex-marriage.

2. *Cherry Hill Seminary*, www.cherryhillseminary.org.
3. Ivo Dominguez, Jr., "Place & Purpose," *Skryclad: Clothed in Visions*, September 12, 2012, http://witchesandpagans.com/EasyBlog/place-purpose.html.
4. M. Macha NightMare, "PantheaCon: Growing Pagan Elders," February 1, 2012, http://machanightmare.com/herself/?ai1ec_event=pantheacon-growing-pagan-elders&instance_id=.
5. Though now twenty years old, this series of articles captures the key disagreements over the issue of professional clergy that remain unresolved today. Andras Corban Arthen, Isaac Bonewits, Judy Harrow, Oriethyia, and Sam Webster, "Panel Discussion: Pagan Clergy," *FireHeart Magazine* 1988–1993, www.earthspirit.com/fireheart/pclergy1.html.
6. See Anne Hill, *The Baby and the Bathwater: What I Learned About Spirituality, Magic, Community, Ecstasy and Power from 25 Years in Reclaiming* (Bodega Bay, CA: Serpentine Media & Music, 2012) and M. Macha NightMare, "A Co-Founder Withdraws from Reclaiming Tradition," *Broomstick Chronicles*, August 6, 2012, www.besom.blogspot.com/2012/08/a-co-founder-withdraws-from-reclaiming.html.
7. Sherry Turkle, *Alone Together: Why We Expect More from Technology and Less from Each Other* (New York: Basic Books, 2011) 164–170.
8. Turkle 199–200.
9. Turkle 201.
10. Turkle 188, 190, 204.
11. Turkle 189–190.
12. Turkle 8–9, 104–105, 109–112.
13. Turkle 116–119.
14. Turkle 290–292.
15. See, for example, Claudine Zap, "Mom sues police and neighbor after she is arrested for letting her kids play outside," *Yahoo! News*, September 19, 2012, http://news.yahoo.com/blogs/sideshow/mom-sues-polices-she-arrested-letting-her-kids-134628018.html. Originally reported by KPRC Channel 2, Houston. See also Lenore Skenazy, "Outrage of the Week: Mom JAILED for Letting Kids Play at Park," FreeRangeKids.com, June 10, 2012, www.freerangekids.com/outrage-of-the-week-mom-jailed-for-letting-kids-play-at-park/. Skenazy began the "Free Range Kids" blog after her article about allowing her nine-year-old to ride the subway alone sparked media furor.
16. David Abram, *The Spell of the Sensuous: Perception and Language in a More-Than-Human World* (New York: Pantheon, 1996) x.
17. Abram ix.
18. Abram 25–26.
19. Abram 43.
20. Abram 45.
21. Abram 52.
22. Abram 68–69.
23. Abram 66, citing Maurice Merleau-Ponty, *The Visible and the Invisible*, trans. Alphonso Lingis (Evanston, IL: Northwestern University Press, 1968).
24. Constance Wise, *Hidden Circles in the Web: Feminist Wicca, Occult Knowledge, and Process Thought* (Lanham, MD: Altamira Press, 2008) 78.
25. Wise 77, quoting Melissa Raphael, *Introducing Thealogy: Discourse on the Goddess* (New York: Continuum, 1999) 146.
26. Wise 78, referring to Carol P. Christ, *Rebirth of the Goddess: Finding Meaning in Feminist Spirituality* (New York: Routledge, 1998) 27.

Bibliography

Abernathey, Marti. "International Transgender Day of Remembrance: Statistics and Other Info." Last updated November 17, 2012. Accessed November 30, 2012. www.transgenderdor.org/statistics.

Abram, David. *The Spell of the Sensuous: Perception and Language in a More-Than-Human World.* New York: Pantheon, 1996.

Aburrow, Yvonne. "Is It Meaningful to Speak of 'Queer Spirituality?' An Examination of Queer and LGBT Imagery and Themes in Contemporary Paganism and Christianity." In *Contemporary Christianity and LGBT Sexualities.* Ed. Stephen Hunt. Burlington, VT: Ashgate, 2009. 139–156.

Albanese, Catherine L. *Nature Religion in America.* Chicago, IL: University of Chicago Press, 1991.

Alexander, Brian. "One Preacher's Message: Have Hotter Sex." *America Unzipped* on NBCNews.com, December 4, 2006. Accessed November 24, 2012. www.msnbc.msn.com/id/13834042/.

Alexander, Brian. "Incurable Gonorrhea May Be Next Superbug." *Sexploration* on MSNBC.com, April 8, 2010. Accessed November 24, 2012. www.msnbc.msn.com/id/36229547/ns/health-sexual_health/.

Althaus-Reid, Marcella. *The Queer God.* New York: Routledge, 2003.

Althaus-Reid, Marcella. "The Church as Dis/Grace." Paper presented at "Queering the Church: Changing Ecclesial Structures" at the Center for Practical Theology, Boston University School of Theology, Boston, MA, April 19, 2007.

American Psychiatric Association. *Diagnostic and Statistical Manual of Mental Disorder.* Fourth Edition, Text Revision. Washington, DC: American Psychiatric Association, 2000.

Anapol, Deborah M. *Polyamory: The New Love without Limits.* San Rafael, CA: IntiNet Resource Center, 1997.

Arthen, Andras Corban, Isaac Bonewits, Judy Harrow, Oriethyia, and Sam Webster. "Panel Discussion: Pagan Clergy." Accessed November 25, 2012. www.earthspirit.com/fireheart/pclergy1.html.

Association of Certified Sexological Bodyworkers. Accessed April 1, 2013. http://sexological bodyworkers.org/.

Attorney General's Commission on Pornography, Final Report. Vol. 1. United States: Department of Justice, July 1986.

Australian Associated Press. "Kinky? You Can't Beat It." *The Sydney Morning Herald*, April 16, 2007. Accessed November 19, 2012. www.smh.com.au/news/national/kinky-you-cant-beat-it/2007/04/16/1176696736407.html.

Baird, Woody. "Pagan Mother Battles for Custody." PolyamorySociety.org. Reprinted from *Yahoo! News*, January 11, 1999. Accessed November 27, 2012. www.polyamorysociety.org/Yahoo-Divilbliss_Article.html.

Benjamin, Ben E. and Cherie Sohnen-Moe. *The Ethics of Touch*. Tucson, AZ: SMA, 2005.

Berger, Helen A., Evan A. Leach, and Leigh S. Shaffer. *Voices from the Pagan Census: A National Survey of Witches and Neo-Pagans in the United States*. Columbia, SC: University of South Carolina Press, 2003.

Berger, Michele Tracy and Kathleen Guidroz, Eds. *The Intersectional Approach: Transforming the Academy through Race, Class, and Gender*. Chapel Hill, NC: University of North Carolina Press, 2009.

Bornstein, Kate. *Gender Outlaw: On Men, Women, and the Rest of Us*. New York: Vintage Books, 1995.

Bornstein, Kate and S. Bear Bergman, Eds. *Gender Outlaws: The Next Generation*. Berkeley, CA: Seal Press, 2010.

Brown, Candy Gunther. "Touch and American Religions." *Religion Compass* 3/4 (2009): 770–783.

Buber, Martin. *I and Thou*. New York: Scribner, 1958.

Budin, Stephanie Lynn. *The Myth of Sacred Prostitution in Antiquity*. Cambridge: Cambridge University Press, 2008.

Burrus, Virginia and Catherine Keller, Eds. *Toward a Theology of Eros: Transforming Passion at the Limits of Discipline*. New York: Fordham University Press, 2007.

Butler, Judith. *Gender Trouble: Feminism and the Subversion of Identity*. New York: Routledge, 1990.

Califia, Pat. "A Personal View of the History of the Lesbian S/M Community and Movement in San Francisco." In *Coming to Power*. Ed. Samois. Boston, MA: Alyson Publications, 1987. 245–283.

Califia, Pat. *Public Sex: The Culture of Radical Sex*. San Francisco, CA: Cleis Press, 1994.

Califia, Pat. "The Way We Live Now." *Poz Magazine*, May 1999. Accessed November 24, 2012. www.poz.com/articles/214_1517.shtml.

Califia, Pat and Drew Campbell, Eds. *Bitch Goddess: The Spiritual Path of the Dominant Woman*. Emeryville, CA: Greenery Press, 1998.

Califia, Patrick. *Speaking Sex to Power: The Politics of Queer Sex*. San Francisco, CA: Cleis Press, 2002.

Califia, Patrick. "The Sadomasochistic Challenge to the Gay Marriage Paradigm." Paper presented at the Annual Meeting of the American Academy of Religion, Montréal, Quebec, Canada. November 19–22, 2009.

Calvert, Robert Noah. *The History of Massage: An Illustrated Survey*. Rochester, VT: Healing Arts/Inner Traditions, 2002.

Caron, Christina. "Phoenix Goddess Temple Raided as Alleged Brothel." *ABC News*, September 9, 2011. Accessed April 1, 2013. http://abcnews.go.com/US/phoenix-goddess-temple-raided-alleged-brothel/story?id=14481945#.UVnTEhzqnGE.

Centers for Disease Control. "Cervical Cancer Statistics." Accessed November 23, 2010. www.cdc.gov/cancer/cervical/statistics/.

Cho, Margaret. "Forward." In *Yes Means Yes! Visions of Female Sexual Power and a World without Rape*. Eds. Jaclyn Friedman and Jessica Valenti. Berkeley, CA: Seal Press, 2008.

Christ, Carol P. "Why Women Need the Goddess." 1978. *Womanspirit Rising: A Feminist Reader on Religion*. Eds. Carol P. Christ and Judith Plaskow. San Francisco, CA: Harper & Row, 1979. 273–287.

Christ, Carol P. *Rebirth of the Goddess: Finding Meaning in Feminist Spirituality.* New York: Routledge, 1998.
Christ, Carol P. *She Who Changes: Re-imagining the Divine in the World.* New York: Palgrave, 2004.
Christian Nymphos (blog). Accessed 30 November, 2012. http://christiannymphos.org/.
Clark-Flory, Tracy. "Will Marriage Change Gay Love?" *Salon.com*, June 28, 2011. Accessed November 24, 2012. www.salon.com/2011/06/28/gay_marriage_23/.
Clifton, Chas S. *Her Hidden Children: The Rise of Wicca and Paganism in America.* Lanham, MD: AltaMira Press, 2006.
Corinna, Heather. "An Immodest Proposal." In *Yes Means Yes! Visions of Female Sexual Power and a World without Rape.* Eds. Jaclyn Friedman and Jessica Valenti. Berkeley, CA: Seal Press, 2008. 179–192.
Coyle, T. Thorn. *Evolutionary Witchcraft.* New York: Penguin, 2004.
Crowley, Aleister. *The Book of the Law.* Boston, MA: Weiser Books, 1976.
Crowley, Aleister. "The Universe: The 0 = 2 Equation." *Magick without Tears.* 1954. Tucson, AZ: New Falcon, 1991. 52–63. Accessed 18 March, 2013. http://hermetic.com/crowley/magick-without-tears/mwt_05.html.
Davis, Phyllis K. *The Power of Touch.* Revised Edition. Carlsbad, CA: Hay House, 1999.
DeCosta-Willis, Miriam. "Introduction." In *Erotique Noire: Black Erotic.* Eds. Miriam deCosta-Willis, Reginald Martin, and Roseann P. Bell. New York: Anchor Books, 1992.
De Grandis, Francesa. *Be A Goddess!* San Francisco, CA: HarperSanFrancisco, 1998.
Divilbiss, April. "PolyFamily Child Custody Case Ends After 2 Year Battle," 1998. Polyamory Society.org. Accessed November 27, 2012. www.polyamorysociety.org/Divilbiss_Families_Case_Ends.html.
DiZerega, Gus. "The Case against 'Pagan Clergy' 4.0." *A Pagan's Blog*, February 2012. Accessed November 25, 2012. http://blog.beliefnet.com/apagansblog/2012/02/the-case-against-pagan-clergy-4–0.html.
Dizon, Don S. and Michael L. Krychman. *Questions & Answers about Human Papilloma Virus (HPV).* Sudbury, MA: Jones & Bartlett Publishers, 2011.
Dominguez Jr., Ivo. "Place & Purpose." *Skryclad: Clothed in Visions*, September 12, 2012. Accessed November 25, 2012. http://witchesandpagans.com/EasyBlog/place-purpose.html.
Dunér, David. *The Natural Philosophy of Emanuel Swedenborg.* New York: Springer, 2013.
Dworkin, Andrea. *Intercourse.* New York: Basic Books, 1987.
Dychtwald, Ken. *Bodymind.* 1977. New York: Penguin, 1986.
Easton, Dossie and Janet W. Hardy (writing as Catherine. A Liszt). *The Ethical Slut: A Guide to Infinite Sexual Possibilities.* San Francisco, CA: Greenery Press, 1997.
Easton, Dossie and Janet W. Hardy. *The New Bottoming Book.* Emeryville, CA: Greenery Press, 2001.
Easton, Dossie and Janet W. Hardy. *The New Topping Book.* Emeryville, CA: Greenery Press, 2003.
Easton, Dossie and Janet W. Hardy. *Radical Ecstasy.* Oakland, CA: Greenery Press, 2004.
Easton, Dossie and Janet W. Hardy. *The Ethical Slut: A Practical Guide to Polyamory, Open Relationships & Other Adventures.* Second Edition. Berkeley, CA: Celestial Arts, 2009.
Eisenberg, Nancy and Janet Strayer, Eds. *Empathy and its Development.* New York: Cambridge University Press, 1990.
Eisler, Riane. *The Chalice and the Blade.* San Francisco, CA: HarperOne, 1988.
Eisler, Riane. *Sacred Pleasure: Sex, Myth, and the Politics of the Body.* San Francisco, CA: HarperSanFrancisco, 1996.
Eller, Cynthia. *The Myth of Matriarchal Prehistory: Why an Invented Past Won't Give Women a Future.* Boston, MA: Beacon Press, 2000.

Eller, Cynthia. *Am I A Woman? A Skeptic's Guide to Gender*. Boston, MA: Beacon Press, 2003.

Ellison, Marvin J. *Erotic Justice: A Liberating Ethic of Sexuality*. Louisville, KY: Westminster John Knox Press, 1996.

"Facts about Violence." Feminist.com. 2008. Accessed November 25, 2012. www.feminist.com/antiviolence/facts.html.

Field, Tiffany. "American Adolescents Touch Each Other Less and Are More Aggressive Toward Their Peers as Compared with French Adolescents." *Adolescence* 34:136 (Winter 1999): 753–758.

Field, Tiffany. *Touch*. Cambridge, MA: MIT Press, 2003.

Foster, Lawrence. *Religion and Sexuality: The Shakers, the Mormons, and the Oneida Community*. Champaign, IL: University of Illinois Press, 1984.

Foster, Lawrence. "Free Love and Community: John Humphrey Noyes and the Oneida Perfectionists." In *America's Communal Utopias*. Ed. Donald E. Pitzer. Chapel Hill, NC: University of North Carolina Press, 1997. 253–278.

Foucault, Michel. *The History of Sexuality: An Introduction*. Vol. I. Trans. Robert Hurley. 1978. New York: Vintage Books, 1990.

The Foundation for AIDS Research, "Assessing the Efficacy of Abstinence-Only Programs for HIV Prevention among Young People," Issue Brief No. 2. Revised October 2007. Accessed June 14, 2013. http://thehill.com/images/stories/whitepapers/pdf/AbstinenceIB.pdf.

Fowles, Stacey May. "The Fantasy of Acceptable 'Non-Consent': Why the Female Sexual Submissive Scares Us (and Why She Shouldn't)." In *Yes Means Yes! Visions of Female Sexual Power and a World without Rape*. Eds. Jaclyn Friedman and Jessica Valenti. Berkeley, CA: Seal Press, 2008. 117–126.

Fox, Matthew. *Original Blessing*. Santa Fe, NM: Bear & Co., 1983.

Friedman, Ellen. *As Above, So Below: A System of Value-Based Ethics for Wiccan Clergy*. Seattle, WA: City University, 2011.

Fuller, Robert C. *Spirituality in the Flesh: Bodily Sources of Religious Experience*. New York: Oxford University Press, 2008.

Gifford, Bob. "The Virtues of Evangelical Sex." *Cogito, Ergo Dubito*, January 25, 2007. Accessed November 24, 2012. http://ergodubito.org/?p=568.

Gilgoff, Dan. "Catholic Church's Sex Abuse Scandal Goes Global." *CNN World*, March 19, 2010. Accessed November 27, 2012. www.cnn.com/2010/WORLD/europe/03/19/catholic.church.abuse/index.html.

Gleick, James. *Chaos: Making a New Science*. New York: Viking, 1987.

Gloeckner, Phoebe. *The Diary of a Teenage Girl: An Account in Words and Pictures*. Berkeley, CA: Frog, 2002.

Glucklich, Ariel. *Sacred Pain: Hurting the Body for the Sake of the Soul*. New York: Oxford University Press, 2001.

Good Morning America. "Good Morning America: Oral Sex is the New Goodnight Kiss." YouTube. May 28, 2009. Accessed November 25, 2012. www.youtube.com/watch?v=Yj8p94a81ZE.

Greer, John Michael. "Myth, History, and Pagan Origins." *The Pomegranate: The International Journal of Pagan Studies* 9 (August 1999): 44–50.

Grossman, Dave. *On Killing: The Psychological Cost of Learning to Kill in War and Society*. New York: Back Bay Books, 1996.

Halperin, David. *One Hundred Years of Homosexuality: And Other Essays on Greek Love*. New York: Routledge, 1990.

Harlow, H.F. "The Nature of Love." *The American Psychologist* 13 (1958): 673–685.

Harlow, H.F. and R. Zimmerman. "Affectional Responses in the Infant Monkey." *Science* 130: 3373 (1959): 421–432.

Harrington, Lee. *Sacred Kink: The Eightfold Paths of BDSM and Beyond*. Lynnwood, WA: Mystic Productions, 2009.

Harris, Dan. "Haggard Admits Buying Meth." ABCNews.com, November 3, 2006. Accessed November 24, 2012. http://abcnews.go.com/GMA/story?id=2626067.

Harris, Sarah. "Teachers Ordered Not to Risk Touching Children for Fear of Legal Action." *Daily Mail*, February 25, 2008. Accessed June 14, 2013. www.dailymail.co.uk/news/article-518289/Teachers-ordered-risk-touching-children-fear-legal-action.html.

Hartshorne, Charles. *Omnipotence and Other Theological Mistakes*. Albany, NY: State University of New York Press, 1984.

"Head Injuries in Football." *New York Times*, October 21, 2010. Accessed November 25, 2012. www.nytimes.com/info/concussions-in-football/.

Hedley, Gil. *Reconceiving My Body*. Xlibris, Self pub., 2000.

Herman, Judith. *Trauma and Recovery*. New York: Basic Books, 1997.

Heyward, Carter. *Touching Our Strength: The Erotic as Power and the Love of God*. San Francisco, CA: Harper & Row, 1989.

Hill, Anne. *The Baby and the Bathwater: What I Learned About Spirituality, Magic, Community, Ecstasy and Power from 25 Years in Reclaiming*. Bodega Bay, CA: Serpentine Media & Music, 2012.

Hoff, Gabriele and Richard A. Sprott. "Therapy Experiences of Clients with BDSM Sexualities: Listening to a Stigmatized Sexuality." *Electronic Journal of Human Sexuality* 12 (September 20, 2009). Accessed November 19, 2012. www.ejhs.org/Volume12/bdsm.htm.

Hopman, Ellen Evert and Lawrence Bond, Eds. *People of the Earth: The New Pagans Speak Out*. Rochester, VT: Destiny Books, 1996.

Hunter, Jennifer. *Rites of Pleasure: Sexuality in Wicca and NeoPaganism*. New York: Citadel Press, 2004.

Hutchins, Loraine. "Bisexual Women as Emblematic Sexual Healers and the Problematics of the Embodied Sacred Whore." *Journal of Bisexuality* 10 (2010): 191–207.

Hutchins, Loraine. "Playing with Sacred Fire: Building Erotic Communities." In *Becoming Visible: Counseling Bisexuals across the Lifespan*. Ed. Beth Firestein. New York: Columbia University Press, 2006. 336–357.

Hutton, Ronald. *The Triumph of the Moon: A History of Modern Pagan Witchcraft*. Oxford: Oxford University Press, 1999.

Hutton, Ronald. *Witches, Druids, and King Arthur*. London: Hambledon & London, 2003.

International Transgender Day of Remembrance. "Statistics and other info." Accessed November 27, 2012. www.transgenderdor.org/statistics.

Johnson, Ramon. "Interview with Gary Gates of The Williams Institute." Accessed November 18, 2012. http://gaylife.about.com/od/index/a/garygates.htm.

Juhan, Deane. *Touched by the Goddess: The Physical, Psychological, and Spiritual Powers of Bodywork*. Barrytown, NY: Station Hill of Barrytown, 2002.

Juhan, Deane. *Job's Body: A Handbook for Bodywork*. Third Edition. Barrytown, NY: Station Hill of Barrytown, 2003.

Jung, C.G. *Civilization in Transition: Volume 10 of the Collected Works of C.G. Jung*. Second Edition. New York: Bollingen, 1970.

Jungkeit, Steve. "Tell-Tale Visions: The Erotic Theology of Craig Thompson's Blankets." In *Graven Images: Religion in Comic Books and Graphic Novels*. Eds. A. David Lewis and Christine Hoff Kraemer. New York: Continuum International Publishing Group, 2010. 333–354.

Kaldera, Raven. *Pagan Polyamory: Becoming a Tribe of Hearts.* Woodbury, MN: Llewellyn Publications, 2005.

Kaldera, Raven. *Dark Moon Rising: Pagan BDSM and the Ordeal Path.* Hubbardston, MA: Asphodel Press, 2006.

Kaldera, Raven. *Hermaphrodeities.* Second Edition. Hubbardston, MA: Asphodel Press, 2009.

Kaldera, Raven. *Power Circuits: Polyamory in a Power Dynamic.* Hubbardston, MA: Alfred Press, 2010.

Kaldera, Raven, Ed. *Sacred Power, Holy Surrender: Living a Spiritual Power Dynamic.* Hubbardston, MA: Alfred Press, 2010.

Karlen, Amo. *Threesomes: Studies in Sex, Power, and Intimacy.* New York: Beech Tree Books, 1998.

Kendall, Todd. "Pornography, Rape, and the Internet." Clemson University, July 2007. Accessed November 27, 2012. www.toddkendall.net/internetcrime.pdf.

Kennelly, Stacey. "Educating for Empathy." *Greater Good: The Science of a Meaningful Life,* July 18, 2012. Accessed December 1, 2012. http://greatergood.berkeley.edu/article/item/educating_for_empathy.

Kincaid, James. *Erotic Innocence: The Culture of Child Molesting.* Durham, NC: Duke University Press, 1998.

Klein, Marty and Charles Moser. "SM (Sadomasochistic) Interests as an Issue in a Child Custody Proceeding." *Journal of Homosexuality* 50: 2 (2006): 233–242. Accessed November 27, 2012. http://sexresearch.com/SSSS/PDFs/20060502_Klein_Moser.pdf. DOI: 10.1300/J082v50n02_11.

Knoploh, Sarah. "ABC Exposes Prevalence of Oral Sex among Teen Girls." *Culture and Media Institute,* May 28, 2009. Accessed November 25, 2012. www.mrc.org/node/28667.

Kraemer, Christine Hoff. *The Erotic Fringe: Sexual Minorities and Religion in Contemporary American Literature and Film.* Ph.D. Boston University, 2008.

Kraemer, Christine Hoff. "Contemporary Paganism, Utopian Reading Communities, and Sacred Nonmonogamy: The Religious Impact of Heinlein and Starhawk's Fiction." *The Pomegranate: The International Journal of Pagan Studies* 13: 1 (2011): 52–76.

Kraemer, Christine Hoff. "Gender and Sexuality in Contemporary Paganism." *Religion Compass* 6: 8 (2012): 390–401.

Kraemer, Christine Hoff. *Seeking the Mystery: An Introduction to Pagan Theologies.* Englewood, CO: Patheos Press, 2012.

Kraemer, Christine Hoff. "The Undying Fire: Divine Love and the Erotic in *Promethea.*" In *Sexual Ideology in the Works of Alan Moore.* Eds. Todd Comer and Joseph Sommers. Jefferson, NC: McFarland, 2012. 150–162.

Kraemer, Christine Hoff and J. Lawton Winslade. "The Magic Circus of the Mind: Alan Moore's *Promethea* and the Transformation of Consciousness through Comics." In *Graven Images: Religion in Comic Books and Graphic Novels.* Eds. A. David Lewis and Christine Hoff Kraemer. New York: Continuum, 2010. 274–291.

Krueger, Richard B. "The DSM Diagnostic Criteria for Sexual Masochism." *Archives of Sexual Behavior.* American Psychological Association, 2010. Accessed January 5, 2013. www.dsm5.org/Research/Documents/Krueger_ASB%20Feb%202011.pdf.

Lauterstein, David. *The Deep Massage Book.* Taos, NM: Complimentary Medicine Press, 2011.

LeVay, Simon. *Gay, Straight, and the Reason Why: The Science of Sexual Orientation.* Oxford: Oxford University Press, 2012.

Lightfoot-Klein, Hanny. "The Sexual Experience and Marital Adjustment of Genitally Circumcised and Infibulated Females in the Sudan." *The Journal of Sex Research* 26: 3 (August 1989): 375–392. Accessed November 19, 2012. www.fgmnetwork.org/articles/sexualexperience.php.

Lorde, Audre. *Sister Outsider: Essays and Speeches by Audre Lorde*. Berkeley, CA: Crossing Press, 1984.

Lupus, P. Sufenas Virius. *All-Soul, All-Body, All-Love, All-Power: A Transmythology*. Anacortes, WA: The Red Lotus Library, 2012.

MacKinnon, Catharine. "Are Women Human?" Interview with Stuart Jeffries. *The Guardian*, April 11, 2006. Accessed November 24, 2012. www.guardian.co.uk/world/2006/apr/12/gender.politicsphilosophyandsociety.

Marcus, Steven. *The Other Victorians: A Study of Sexuality and Pornography in Mid-Nineteenth-Century England* (1966). Piscataway, NJ: Transaction Publishers, 2008.

Marcuse, Herbert. *Eros and Civilization: A Philosophical Inquiry into Freud*. Boston, MA: Beacon Press, 1966.

Mattheson, Elise, Ed. "Alt.Polyamory Frequently Asked Questions." *Internet FAQ Archives*. Last modified June 10, 1997. www.faqs.org/faqs/polyamory/faq/.

Melton, J. Gordon. "Reiki: The International Spread of a New Age Healing Movement." *New Age Religion and Globalization*. Ed. Mikael Rothstein. Aarhus: Aarhus University Press, 2001. 73–93.

Merleau-Ponty, Maurice. *The Visible and the Invisible*. Trans. Alphonso Lingis. Evanston, IL: Northwestern University Press, 1968.

Momo, Comfort. "Female Genital Mutilation." In *Female Genital Mutilation*. Ed. Comfort Momo. Abingdon, UK: Radcliffe, 2005. 5–12.

Montagu, Ashley. *Touching: The Human Significance of the Skin*. Third Edition. New York: Harper & Row, 1986.

Moore, Alan and Dave Gibbons (penciller). *Watchmen*. New York: DC Comics, 1987.

Moore, Alan, J.H. Williams (penciller), and Mick Gray (inker). *Promethea* Vol. 4. Reprints issues #19–25. La Jolla, CA: America's Best Comics, 2003.

Moore, Alan and José Villarrubia (illustrator). *The Mirror of Love*. Atlanta, GA: Top Shelf Productions, 2004.

Morey, Ann-Janine. *Religion and Sexuality in American Literature*. Cambridge: Cambridge University Press, 1992.

Myers, Brendan. *The Other Side of Virtue*. Hants, UK: O Books, 2008.

National Highway Traffic Safety Administration. "An Analysis of the Significant Decline in Motor Vehicle Traffic Fatalities in 2008." Last modified June 2010. www-nrd.nhtsa.dot.gov/Pubs/811346.pdf. Document number DOT HS 811 346.

Newcomb, Alyssa. "New Jersey School Bans Hugging." *ABC News*, March 23, 2012. Accessed November 24, 2012. http://abcnews.go.com/blogs/headlines/2012/03/new-jersey-school-bans-hugging/.

Newitz, Annalee. "Love Unlimited: The Polyamorists." *New Scientist* 191: 2559 (8 July 2006): 44–47.

Newman, Maria. "Cautious Teachers Reluctantly Touch Less; A Fear of Abuse Charges Leads to Greater Restraint With Students." *New York Times*, June 24, 2008. Accessed November 25, 2012. www.nytimes.com/1998/06/24/nyregion/cautious-teachers-reluctantly-touch-less-fear-abuse-charges-leads-greater.html.

Nichols, Margaret. "Psychology and BDSM: Pathology or Individual Difference?" *New Jersey Psychologist*, 2002. Accessed November 27, 2012. www.ipgcounseling.com/psychology_and_bdsm.html.

Nielsen, Morten Ebbe Juul. "Safe, Sane, and Consensual—Consent and the Ethics of BDSM." *The International Journal of Applied Philosophy* 24:2 (2010): 265–288.

NightMare, M. Macha. "Reclaiming-Tradition Witchcraft." *Reclaiming Quarterly* 88 (Autumn 2002). Accessed November 25, 2012. www.reclaimingquarterly.org/88/rq-88-tradition.html.

NightMare, M. Macha. "PantheaCon: 'Growing Pagan Elders.'" February 1, 2012. Accessed November 25, 2012. http://machanightmare.com/herself/?ai1ec_event=pantheacon-growing-pagan-elders&instance_id=.

NightMare, M. Macha. "A Co-Founder Withdraws from Reclaiming Tradition." *Broomstick Chronicles*, August 5, 2012. Accessed November 27, 2012. www.besom.blogspot.com/2012/08/a-co-founder-withdraws-from-reclaiming.html.

NightMare, M. Macha. "Back Story, One of Several." *Broomstick Chronicles*, August 8, 2012. Accessed June 14, 2013. www.besom.blogspot.com/2012/08/back-story-one-of-several.html.

OTO Women's Symposium 2006. "Presentations." Accessed November 25, 2012. http://notocon.org/past/otows/i/presentations.html.

Pagels, Elaine. *The Gnostic Gospels*. New York: Vintage Books, 1989.

Pagnozzi, Amy. "Sexual Intolerance Is the Shame in Suicide." *Hartford Courant*, January 29, 2002. Accessed November 27, 2012. http://articles.courant.com/2002–01–29/news/0201291092_1_adult-entertainment-theater-shame.

Paper, Jordan D. *The Deities Are Many: A Polytheistic Theology*. Albany, NY: State University of New York Press, 2005.

Parliament Statement of Indigenous People, *Council for a Parliament of the World's Religions*, December 9, 2009. Accessed June 14, 2013. www.parliamentofreligions.org/news/index.php/2010/07/new-release-pwr-statement-of-indigenous-peoples/.

Peterson, Dale and Richard Wrangham. *Demonic Males: Apes and the Origins of Human Violence*. New York: Houghton Mifflin, 1997.

Pew Forum on Religion and Public Life. *Church Statistics and Religious Affiliations*, 2008. Accessed November 24, 2012. http://religions.pewforum.org/affiliations.

Pike, Sarah. "Gleanings from the Field: Leftover Tales of Grief and Desire." In *Researching Paganisms*. Eds. Jenny Blain, Douglas Ezzy, and Graham Harvey. Walnut Creek, CA: AltaMira Press, 2004. 97–112.

Pitzl-Waters, Jason. "Parsing the Pew Numbers." *The Wild Hunt*, February 26, 2008. Accessed November 24, 2012. http://wildhunt.org/blog/2008/02/parsing-pew-numbers.html.

Pitzl-Waters, Jason. "Interview with Starhawk." *The Wild Hunt*, September 28, 2009. Accessed November 27, 2012. http://wildhunt.org/2009/09/interview-with-starhawk.html.

Prescott, James W. "Body Pleasure and the Origins of Violence." *The Futurist*, April 1975. Reprinted in *The Bulletin of Atomic Scientists* 31 (November 1975): 10–20. Accessed November 27, 2012. www.violence.de/prescott/bulletin/article.html.

Prescott, James W. "Alienation of Affection." *Psychology Today*, December 1979. Reprinted in *The Truth Seeker*, March 1989: 22. Accessed November 27, 2012. www.violence.de/prescott/letters/Alienation_of_Affection.pdf.

Prescott, James W. "Affectional bonding for the prevention of violent behaviors: Neurobiological, psychological and religious/spiritual determinants." In *Violent Behavior: Assessment and Intervention*. Vol. 1. Eds. L.J. Herzberg, G.F. Ostrum, and J. Roberts Field. Great Neck, NY: PMA Publishing, 1990. 110–142.

Press, Bill. "The Sad Legacy of Jerry Falwell." *The Milford Daily News*, May 18, 2007. Accessed November 24, 2012. www.milforddailynews.com/opinion/x1987843539.

Privateer, Jennifer. "Breaking the Boundaries of Sex and Love: An Interview with Madrone." *Reclaiming Quarterly* 82 (Spring 2001). Accessed November 24, 2012. www.reclaimingquarterly.org/82/rq-82-breakingboundaries.html.

Raphael, Melissa. *Introducing Thealogy: Discourse on the Goddess*. New York: Continuum, 1999.

Reich, Wilhelm. *Character-Analysis*. New York: Noonday, 1949.

Reinisch, June M. with Ruth Beasley. *The Kinsey Institute New Report on Sex*. New York: St. Martin's Press, 1990.

Richters, Juliet, Richard O. de Visser, Chris E. Rissel, Andrew E. Grulich, and Anthony M.A. Smith. "Demographic and Psychosocial Features of Participants in Bondage and Discipline, 'Sadomasochism' or Dominance and Submission (BDSM): Data from a National Survey." *Journal of Sexual Medicine* 5 (2008): 1660–1668.

Rubin, Gayle. "Elegy for the Valley of Kings." In *Changing Times: Gay Men and Lesbians Encounter HIV/AIDS*. Eds. Martin P. Levine, Peter M. Nardi, and John H. Gagnon. Chicago, IL: University of Chicago Press, 1997. 101–144.

Rubin, Gayle. *Deviations: A Gayle Rubin Reader*. Durham, NC: Duke University Press, 2011.

Savage, Dan. "Youth Pastor Watch." *The Stranger*, June 28, 2010. Accessed November 30, 2012. http://slog.thestranger.com/slog/archives/2010/06/28/youth-pastor-watch.

Scarry, Elaine. *The Body in Pain: The Making and Unmaking of the World*. New York: Oxford University Press, 1985.

Schnarch, David. *Passionate Marriage: Keeping Love & Intimacy Alive in Committed Relationships*. New York: Henry Holt & Company, 1997.

Sedgwick, Eve Kosofsky. *Tendencies*. Durham, NC: Duke UP, 1993.

Seidman, Steven. "The Power of Desire and the Danger of Pleasure: Victorian Sexuality Re-Considered." *Journal of Social History* 24: 1 (Autumn, 1990): 47–67.

Skenazy, Lenore. "Outrage of the Week: Mom JAILED for Letting Kids Play at Park." FreeRange Kids.com, June 10, 2012. Accessed November 28, 2012. www.freerangekids.com/outrage-of-the-week-mom-jailed-for-letting-kids-play-at-park/.

Spitz, R.A. "Hospitalism: A follow-up report." In *The Psychoanalytic Study of the Child*. Vol. 2, Eds. D. Fenichel, P. Greenacre, and A. Freud. New York: International Universities Press, 1947. 113–117.

Starhawk. *Truth or Dare: Encounters with Power, Authority, and Mystery*. San Francisco, CA: HarperSanFrancisco, 1987.

Starhawk. *The Fifth Sacred Thing*. New York: Bantam Books, 1993.

Starhawk. *The Spiral Dance: A Rebirth of the Ancient Religion of the Great Goddess*. 20th Anniversary Edition. San Francisco, CA: HarperSanFrancisco, 1999.

Strmiska, Michael F. "Modern Paganism in World Cultures: Comparative Perspectives." *Modern Paganism in World Cultures: Comparative Perspectives*. Santa Barbara, CA: ABC-CLIO, 2005. 1–53.

Suttie, Jill. "The Politics of Empathy." *Greater Good: The Science of a Meaningful Life*, June 16, 2010. Accessed December 1, 2012. http://greatergood.berkeley.edu/article/item/the_politics_of_empathy.

Taormino, Tristan. *Opening Up: A Guide to Creating and Sustaining Open Relationships*. San Francisco, CA: Cleis Press, 2008.

The 1928 Book of Common Prayer. New York: Oxford University Press, 2007.

Thompson, Craig. *Blankets*. Marietta, GA: Top Shelf Productions, 2003.

Thompson, Mark, Ed. *Leatherfolk: Radical Sex, People, Politics, and Practice*. Los Angeles, CA: Daedalus Publishing, 2004.

Tolman, Deborah L. *Dilemmas of Desire: Teenage Girls Talk about Sexuality*. Cambridge, MA: Harvard UP, 2002.

Trenholm, Christopher, Barbara Devaney, Ken Fortson, Lisa Quay, Justin Wheeler, and Melissa Clark. "Impacts of Four Title V, Section 510 Abstinence Education Programs, Final Report," Mathematica Policy Research, April 2007. Accessed June 14, 2013. http://aspe.hhs.gov/hsp/abstinence07/index.htm.

Troost, Hazel/Cedar. "Reclaiming Touch: Rape Culture, Explicit Verbal Consent, and Body Sovereignty." In *Yes Means Yes! Visions of Female Sexual Power and a World without Rape*. Eds. Jaclyn Friedman and Jessica Valenti. Berkeley, CA: Seal Press, 2008. 171–178.

Turkle, Sherry. *Alone Together: Why We Expect More from Technology and Less from Each Other.* New York: Basic Books, 2011.

UniteWomen.org. Facebook.com profile. Timeline Photos, November 29, 2012. Accessed November 30, 2012. www.facebook.com/photo.php?fbid=439038836157661&set=a.302468136481399.68437.302448176483395.

Urban, Hugh B. *Magia Sexualis: Sex, Magic, and Liberation in Modern Western Esotericism.* Berkeley, CA: University of California Press, 2006.

Vale, V. and John Sulak (Eds). *Modern Pagans: An Investigation of Contemporary Pagan Practices.* San Francisco, CA: Re/Search, 2001.

Valiente, Doreen. *Charge of the Goddess: The Mother of Modern Witchcraft.* Brighton, UK: Hexagon Hoopix, 2000.

Valiente, Doreen. "The Charge of the Goddess." The Doreen Valiente Foundation. Accessed November 25, 2012. http://doreenvaliente.org/2009/06/poem-the-charge-of-the-goddess/.

Vanderbilt, Shirley. "Eating Disorders: Touch Can Ease the Symptoms." *Massage and Bodywork Magazine* August/September 2001. Accessed November 26, 2012. www.massagetherapy.com/articles/index.php/article_id/552/Eating-Disorders.

Wesley, John. "A Plain Account of Christian Perfection." In *The Works of John Wesley* (1872). Vol. 11. Ed. Thomas Jackson. Grand Rapids, MI: Baker Book House, 1996.

Wesley, John. *Explanatory Notes upon the New Testament* (1754) Forgotten Books, 2012.

WGBH Educational Foundation and Clear Blue Sky Productions. "Chimps and Bonobos," *Evolution*, 2001. Accessed November 27, 2012. www.pbs.org/wgbh/evolution/library/07/3/l_073_03.html.

Wilcox, Melissa M. "Innovation in Exile: Religion and Spirituality in Lesbian, Gay, Bisexual, and Transgender Communities." In *Sexuality and the World's Religions.* Eds. David W. Machacek and Melissa M. Wilcox. Santa Barbara, CA: ABC-CLIO, 2003.

Williams, Brandy. *The Woman Magician: Revisioning Western Metaphysics from a Woman's Perspective.* Woodbury, MN: Llewellyn, 2011.

Winslade, Jason Lawton. "When Pan Met Babalon: Challenging Sex Roles at a Thelemic/Pagan Festival." Pagan Studies Group. Annual Meeting of the American Academy of Religion. Chicago, IL, November 16–20, 2012.

Wise, Constance. *Hidden Circles in the Web: Feminist Wicca, Occult Knowledge, and Process Thought.* Lanham, MD: Altamira Press, 2008.

Wroe, Martin. "Turbulent Priest Ministers to the New Age Soul: Martin Wroe Meets the Friar Delivering 'Pagan Heresy' to Packed Churches." *The Independent*, July 14, 1992. Accessed March 18, 2013. www.independent.co.uk/news/uk/turbulent-priest-ministers-to-the-new-age-soul-martin-wroe-meets-the-friar-delivering-pagan-heresy-to-packed-churches-1533115.html.

York, Michael. *Pagan Theology.* New York: New York University Press, 2003.

Yronwode, Nagasiva, "V.H.," C. Baphemetis, Catherine Yronwode, and Tom Worrel. "Was Aleister Crowley A Sexist?" *The Arcane Archive*, 1999–2000. Accessed November 25, 2012. http://arcane-archive.org/faqs/crowleysexistfaq.php#sexist.

Zap, Claudine. "Mom Sues Police and Neighbor After she is Arrested for Letting her Kids Play Outside." *Yahoo! News*, September 19, 2012. Originally reported by KPRC Channel 2, Houston. Accessed November 28, 2012. http://news.yahoo.com/blogs/sideshow/mom-sues-polices-she-arrested-letting-her-kids-134628018.html.

Zell, Morning Glory. "A Bouquet of Lovers: Strategies for Responsible Open Relationships." *Green Egg* 89 (Beltane 1990). Reprinted on "CAWeb—Home of the Church of All Worlds." Accessed November 18, 2012. http://caw.org/content/?q=bouquet.

Acknowledgments for Previously Published Material

The author and publisher would like to gratefully acknowledge the copyright holders for permission to reprint material:

Kraemer, Christine Hoff. Revised selections from *Seeking the Mystery: An Introduction to Pagan Theologies*. © 2012 by Christine Hoff Kraemer. By permission of Patheos Press, Englewood, CO.

Kraemer, Christine Hoff, revised selections from "The Undying Fire: Divine Love and the Erotic in *Promethea*." From *Sexual Ideology in the Works of Alan Moore: Critical Essays on the Graphic Novels*. © 2012 Edited by Todd A. Comer and Joseph Michael Sommers. By permission of McFarland & Company, Box 611, Jefferson NC 28640.

Simos, Miriam (Starhawk). From *The Fifth Sacred Thing*. Copyright © 1993 by Miriam Simos. Used by permission of Bantam Books, a division of Random House. Any third party use of this material, outside of this publication, is prohibited. Interested parties must apply directly to Random House for permission.

Index